SHAMROCKS AND OIL SLICKS

Shamrocks and Oil Slicks

A People's Uprising Against Shell Oil in County Mayo, Ireland

by FRED A. WILCOX

MONTHLY REVIEW PRESS
New York

Copyright © 2019 Fred A. Wilcox
All Rights Reserved

Library of Congress Cataloging-in-Publication Data available from the publisher

ISBN 978-158367-846-6 pbk
ISBN 978-158367-847-3 cloth

Typeset in Minion Pro

MONTHLY REVIEW PRESS, NEW YORK
monthlyreview.org

5 4 3 2 1

Contents

Preface | 7

Introduction | 9

Farmer | 17

Fisherman | 31

Witness | 45

Scholar | 55

Communicator | 71

Innkeepers | 87

Musicians | 101

Journalist | 117

Teacher | 127

Orders | 143

Acknowledgments | 156

Glossary | 157

Notes | 163

Related Reading | 167

This book is dedicated to Brendan Benjamin Wilcox, writer, photographer, artist, father, and altogether beautiful soul. He loved his friends and family, and left us to grieve his passing.

We declare the right of the people of Ireland to the ownership of Ireland, and to the unfettered control of the Irish destinies, to be sovereign and indefeasible.

—Proclamation of Poblacht na hÉireann, Provisional Government of the Irish Republic, 1916

Preface

IN 1970, THE WAR IN VIETNAM was raging. American cities were burning. Assassins had killed John F. Kennedy, his brother Bobby, the Reverend Dr. Martin Luther King Jr., civil rights workers, Black Panthers, and others. The Ohio National Guard opened fire on anti-war protestors at Kent State, killing four students. During an uprising at Attica state prison, police shot and killed more than forty inmates and guards.

Sick at heart, convinced the United States had gone insane, I flew to Ireland. In secondary school, we'd read books by English writers, learned English history. Our teachers did not talk about imperialism or colonialization. We were not told about Ireland's long struggle to become a free, democratic nation.

I wandered from town to town, meeting people in pubs, joining singalongs of rebel songs, standing alone at night beside the Atlantic Ocean, feeling as though I'd come home. I rode a freight train into Belfast, arriving at the height of the Troubles, British troops racing through the streets in armored cars, smoldering neighborhoods, a drunken British soldier waving his rifle in my face.

In 1975 I returned to Ireland, living for a time on Inishbofin, "island of the white cow," where I read Irish history, studied Irish literature, and learned about the Celts. Years later I would serve with a group of international observers in Northern Ireland during the marching season, a time when pro-British, anti-Catholic groups parade through towns and cities. One summer I acted as spokesperson for the observers. Caught in a Belfast riot, bricks flying overhead, cars exploding close by, the police firing hard rubber bullets, water cannons knocking over protestors, I experienced firsthand the rage of people who'd been forced to live in substandard housing, to work at low-paying jobs, and to suffer for decades from endless sectarian violence.

My son, Brendan B. Wilcox, and I spent a lot of time in Ireland, raising pints of Guinness Stout, listening to music, searching for good food. I facilitated writers' workshops on Inisheer, one of the Aran Islands off the coast of Galway, and in Tralee, County Kerry.

Brendan loved Ireland, the Irish people loved him. We talked often about moving there.

In 2014 I read about a situation in County Mayo, where local people were resisting Shell Oil's plans to build a gas refinery and a dangerous pipeline. I did not think government officials would allow this project to go forward: Tens of millions of tourists visit Ireland each year, not to see a gas refinery or to worry about a pipeline exploding close to the hotel, hostel, or bed-and-breakfast in which they are staying.

It turns out that I had not learned very much about the country I'd adopted as my own.

Writing *Shamrocks and Oil Slicks*, I encountered the soul sickness of greed and corruption, and also the meaning of great courage. Some of the people I met, and with whom I became friends, feel that Ireland needs revolution. I could not agree more.

—FRED A. WILCOX
JULY 28, 2019

Introduction

VISITORS TO COUNTY MAYO, IRELAND, enjoy hiking on misty mountain trails, sailing on cold clear lakes, fishing in rivers teeming with salmon, and picnicking on pristine ocean beaches. Whales frolic in tranquil bays, rare birds nest in seaside cliffs, and sheep meander, as though deep in thought, across quiet roads.

In the sixteenth century, pirate queen Grace O'Malley sailed from here to plunder the British fleet. St. Patrick is reputed to have taught Christianity in Mayo. Native Michal Davitt (l846–1906) cofounded the National Land League in Mayo, devoting his life to helping tenants break the power of landlords.

John Synge set his plays *Riders to the Sea* and *Playboy of the Western World* in the wilds of Mayo. John Ford filmed *The Quiet Man* in the village of Cong on the Mayo-Galway border. A half-century later, fans pose for photographs in costumes worn by the film's stars, John Wayne, Maureen O'Hara, and Barry Fitzgerald.

In County Mayo, say people who live here, the land speaks to those who care to listen.

IN OCTOBER 1996 ENTERPRISE OIL Ireland discovers a 250 million-year-old gas deposit 83 kilometers (52 miles) off Mayo's coast.

A trillion to 850 billion cubic feet of gas lies in this, the Corrib gas field. Gas for cooking meals and heating homes, for stimulating Ireland's economy, and, claims EOI, improving the lives of Irish citizens.

By 2002, having acquired primary interest in the Corrib gas deposits, Royal Dutch Shell Oil initiates a plan to build the first refinery on land in Ireland. Shell's engineers, say company spokespersons, are the best in the world. *The project will bring new jobs, improved roads, money for schools, and more to this impoverished area. County Mayo's environment will never be harmed. Tourists will always visit this magical region.*

Moreover, this multinational corporation is committed to being a good neighbor, to listening to the community, and to resolving differences through constructive conversation.[1]

Clergy call Shell's new project a "godsend." There will be no more sorrowful goodbyes in Mayo when friends and relatives are forced to leave home in search of employment. Never again will families have to struggle to put food on the table and to keep a roof over their children's heads. Happy days. Everyone will soon be rich.

Shell intends to construct a pipeline from where the gas comes ashore nine kilometers (approximately six miles) inland to the village of Ballinaboy. The pipe will be buried in unstable soil, run through farmers' fields, and be dangerously close to private homes. Pressure inside the pipeline will be so high that in the event of an accident, families will be incinerated in seconds.

A MAN KNOCKS ON WILLIE CORDUFF'S door. He is there to take measures and dig holes for the gas pipeline that will run beneath Willie's fields. Not to worry, he says, the digger he'll use won't disturb anything. Corduff and his family have spent a lifetime working to reclaim land from the bog. This is their home, their sense of place, where they belong. No one will ever change that.

Willie orders the man to leave. Accompanied by the police, he soon returns.

Willie knows that Shell Oil has constructed pipelines and refineries all over the world. The company doesn't need to explain its plans to "begrudgers who think they know more about high-tech projects than do professional engineers." After all, Shell is providing jobs. Other people in this backwater are feeling optimistic. Small-minded people like himself always resist progress. Complain all they want, but this pipeline and refinery *will be built.*

A new invader has arrived in Mayo. Not a landlord's vicious agents or uniformed men with guns, but a powerful corporation with legions of well-paid lawyers, professional engineers, high-powered executives, and public relations staff skilled in the art of convivial deceit:

> Shell's external relations team in Mayo is dedicated to engaging and working with the local community in order to maximize the benefits and minimize the impacts of the Corrib gas facilities. [Shell] is committed to being a good neighbor and to listening and addressing community concerns. We also have a detailed complaints process in place, which is managed by our community liaison team.[2]

LOCAL PEOPLE TRY TO CONVEY their concerns to Shell. *Why risk injuring and killing people by constructing a refinery so close to a village? Why not construct it in some desolate, unpopulated area?*

Some time later, they will joke about their naïveté.

"None of us had ever been in a courtroom," they will say. "Didn't know what a judge looked like. We respected the police and trusted the government. We knew very little about companies like Shell."

Property owners find intruders digging holes and taking measurements on their land. Asked to leave, these strangers insist they have permits to do this work. Permits? Yes, from Shell Oil. Unfortunately, they always seem to have left these permits back at the office.

The state "is bound to protect the personal rights of the citizen, and in particular to defend the life, person, good name, and property rights of every citizen."³

When farmers block Shell's access to their fields, the Minister for Ireland's Marine and Natural Resources issues a *Compulsory Acquisition Order*, denying citizens the right to keep Shell off their property. Crews arrive with police escorts.

Farmers, fishermen, housewives, and others begin a nonviolent campaign to resist construction of the refinery. Police punch and kick demonstrators, beat them with batons, and toss them into ditches. Shell hires mercenaries who harass, intimidate, and assault its opponents. Media outlets call resisters "terrorists" and accuse them of belonging to the Irish Republican Army (IRA).

"THE EARTH," WRITES POPE FRANCIS in his *Encyclical on Climate Change and Equality*, "our home, is beginning to look more and more like an immense pile of filth. In many parts of the planet, the elderly lament that once beautiful landscapes are now covered with rubbish. . . . Frequently no measures are taken until after people's health has been irreversibly affected."⁴

IN 2005, A HIGH COURT JUDGE sends one schoolteacher and four property owners who'd denied Shell access to their land—they would become known as the Rossport 5—to indefinite terms in prison. If this conflict continues, someone might get seriously injured, or even killed.

Shell Oil is establishing a beachhead in County Mayo, preparing the way for fossil fuel companies to build refineries near ancient villages, and to lace the countryside with elaborate patchworks of dangerous pipelines.

Ireland is not Nigeria, where a dictatorship supported by Shell turned the Ogoni people's land into one vast poisonous oil slick. A country where, during the 1990s, the military government executed environmentalists and slaughtered thousands of men, women, and children in order to protect multinational corporations.

Ireland is presumably a democracy, with a constitution that

articulates citizens' rights. Its army does not take orders from a dictator to torture, jail, and kill its own citizens. Ireland's police, the An Garda Síochána, do not routinely rob, rape, disappear, and massacre their own people.

Still, a rich and powerful multinational corporation was allowed to ignore rights guaranteed to the Irish people, using violence against peaceful demonstrators, disrupting communities, and traumatizing citizens who are determined to protect the environment they love.

AT MIDNIGHT ON JANUARY 1, 2015, the skies over Ballinaboy, County Mayo, explode with light. A huge fiery middle finger thrusts at those who had waged a prolonged struggle to keep Shell from building its refinery: a clear message to anyone who might dare to challenge this corporation's right to do whatever it wishes, wherever it wishes, to whomever it wishes.

The Corrib oil refinery and pipeline, against which so many people had expended so much energy, time, and tears, was in full operation.

Multinational companies use bribes, mercenaries, assassins, armies, and corrupted police and legislatures and courts to secure and exploit the world's natural resources. They pay pseudo-scientists to promote the fiction that human activity is not responsible for global warming. They build pipelines that leak and explode, killing and maiming people. They float tankers that seep crude oil onto beaches and wetlands, destroying the habitats of birds and mammals and fishes, reptiles and amphibians and smaller organisms, and leaving swaths of unemployed, destitute people.

Nation-states enshrine laws to protect endangered environments. World bodies establish international agreements to preserve the planet's most vulnerable regions. Still, avaricious corporations, colluding with corrupt politicians and "civil servants," are determined to cut down the last tree in the last rain forest; to poison the last clean water; destroy the last coral reef; make extinct the last protected animals and sea life.

NEVERTHELESS, THE MOVEMENT TO create a world powered by sun and wind and water is growing. Uprisings are breaking out all over the planet, many led by people under the age of 21. People know that fossil fuel companies are dinosaurs, destined to go extinct. When this happens, pathologists will conclude that these behemoths overdosed on greed—the planet's most addictive drug.

A fierce conflict is raging between those who see the world as one vast exploitable resource belonging exclusively to them, and those who believe human beings must live in harmony with trees, plants, and animals.

According to Global Witness, an organization that campaigns to end environmental and human rights abuses, more than three people were killed *each week* in 2015, and two hundred people were killed in 2016, trying to defend their land, forests, and rivers against destructive industries:

> "As demands for products like minerals, timber and palm oil continue," says Global Witness campaign leader Billy Kyte, "governments, companies and criminal gangs are seizing land in defiance of the people who live on it. Communities that take a stand are increasingly finding themselves in the firing line of companies' private security, state forces and a thriving market for contract killers. For every killing we document, many others go unreported. Governments must urgently intervene to stop this spiraling violence."[5]

Those who insist that they have the right to buy and sell all living things dismiss warnings that we must act to save the natural world. After all, say multinational corporations, there's no limit to our planet's resources. It's impossible, they say, to open too many coal mines, to build too many pipelines, or construct too many refineries and chemical plants. It's not possible to poison the world's great rivers, pollute the oceans, poison the food and water humans and animals need in order to survive.

Two thousand eminent scientists warn that unless we greatly reduce carbon emissions now, coastal cities will go under water. Small fishing villages like the ones that ring Ireland will be wiped out. Hundreds of millions of people throughout the world will be forced to flee from the devastating effects of droughts, violent storms, and floods.

Shamrocks and Oil Slicks is the story of people whose love for the sea, their rivers and lakes and bogs, their friends and families, their heritage, inspired them to wage a relentless fifteen-year nonviolent struggle against one of the world's most destructive predators.

Abandoned by politicians, maligned by the media, beaten by mercenaries and the police, sent to prison, they chose truth over lies, courage over cowardice, and life over death. They represent the future—a sustainable world in which all creatures celebrate the gift of living together on the only planet on which we know, for certain, that life exists.

Willie and Mary Corduff on their farm

Farmer

> I urgently appeal, then, for a new dialogue about how we are shaping the future of our planet. We need a conversation which includes everyone since the environmental challenge we are undergoing, and its human roots, concern and affect us all.
> —Pope Francis, Laudato Si', mi' Signore: On Care for Our Common Home (Encyclical, 2015)

"They couldn't understand why I was laughing. I'm not for sale.

"'Oh come now, Willie,' my solicitors said. 'Tell us. How much would you take. Name your price, Willie. It could be a million.'

"'Shell hasn't got the money.'

"They kept right on talking. 'Don't be daft, Willie,' they said. 'They've more money than you'd ever know. Take the money and you can fight them in the courts.'

"They couldn't understand 'Name your price.' 'I'm not for sale.'

"Then people started making up stories: 'Willie Corduff is just waiting for more money. That's why he's holding back for so long. More money.'"

Solicitors urged Willie's wife, Mary, to talk some sense into her husband's head, urging her to take the money.

" 'You can get a better farm.' "

"My heart and soul was here since I was a kid," says Willie. "How could I go somewhere that would mean nothing to me?"

"That was the killing point," adds Mary. "Our own solicitors. They were serious; they meant business, kept telling me to talk Willie into it."

WILLIE CORDUFF HAS LIVED ALL of his life on a small County Mayo farm beside the sea, a fourth generation of Corduffs; his own father and grandfather were here when everyone was poor and a collective memory of the Great Hunger (the deadly Irish "famine" that began in 1845) held sway over the land and its people. Survival, they knew from hard experience, depends on cooperation between friends, neighbors, and family. There were no supermarkets or shopping malls back then. Families planted vegetable gardens, sold turf, and took their farm animals to market to sell or trade. People bartered, shared what little they had.

Willie's parents and their five children lived with a timeless sense of history, working on land the way their ancestors had, traveling to town once a month but going no farther. Things did change, loved ones got sick and died, others left for work abroad, but the bonds of this ancient world seemed unbreakable. Folks knew their neighbors, counted on them for help. Willie's own father would climb out of bed late at night to help someone down the road whose cow was calving or whose sheep might be lambing.

During my first trip to Mayo in 2014, I did not meet Willie and Mary, but later I watched Terence Conway's video footage of Willie facing off with the police at demonstrations against Shell Oil's project. Holding out his arms, roaring, "Do your duty now. If I'm breaking the law, arrest me. Go ahead, put the cuffs on my wrists. Arrest me."

Conway's videos show police beating people at protests, but arresting few.

IT'S NOW SEPTEMBER 2017. We arrive midmorning at Willie and Mary's home. Two sheepdogs greet us like old friends, sun-dappled fields slope to the blue-green sea, the soft quiet of northwestern Ireland.

The land on which this modest home sits was originally bog, reclaimed by backbreaking labor. As a boy, Willie worked with his father constantly to improve their farmlands. Then, just a few years after Mary and Willie were married, fire destroyed their outside shed, leaving them in debt and with few prospects for earning money.

Willie worked on the roads for a time, but he needed a car to keep that going. With six children to support, he cut turf and seaweed. He fished, grew potatoes, raised milk cows. Difficult times, says Mary, that made the family stronger, teaching them good values and gratitude for life.

Seated at his kitchen table, Willie appears smaller than he does in video footage of protests, with a warm smile and the aura of a man who, having survived terrible beatings, months in prison, fear, betrayal, and disillusionment, treasures a cup of tea on a warm fall day.

Mary and Willie Corduff paid a big price for refusing to go along with Shell Oil's schemes. Nevertheless, they are warm and hospitable, welcoming Terence Conway and me with tea, coffee, and cookies; later she'll prepare a delicious lunch of homemade soup and sandwiches.

IT'S 2002. AN EMPLOYEE OF SHELL OIL, then primary owner of the Corrib gas project, stops by the Corduff home. Not, recalls Willie, to talk to the family or ask questions. Not to request permission to walk about the farm.

No prior letter or phone call, just this man knocking at the door, announcing that the gas company is going to run a pipeline beneath Mary's clotheslines, past laughing children, grazing cows and horses.

Willie remembers for us: "This man, with a Scottish accent,

wants to show me something. 'Why I'm here,' he says, 'is to show you where we'll be digging trial holes.'"

Willie adds an aside to us: "Now, I love the land. I adore the land.

"'Right in there,' the man says. 'We'll be using a JCB, Willie. Do you know what that is? It's a machine that you let the legs down and it won't do any damage.'

"'How come it will do no damage?' I ask. This goes on for a while, and then he starts to cop on, and I say, 'About one thing you are right.'

"'What's that?' he says. 'I will guarantee there will be no damage, that's what.'

"I point to a pitchfork leaning against a wall, and I say, 'Now, do you know what that is?'

"'I don't,' he says.

"So I explain. 'That thing will do some real damage. And the best thing for you is to get out of here fast as you can.'

"He tries to tell me that my neighbors have already signed, giving Shell permission to come on to their land.

"We see our neighbor later and ask if he's met Shell's man, and he says, 'Oh no, not at all.'

"They started out with divide and conquer, lying to us right from the beginning."

WILLIE AND MARY SOON REALIZE that Shell isn't interested in Ireland's turbulent history, its great writers and artists and saints, its many uprisings against tyranny. This company does not care about ordinary people who toil to make land productive and to preserve their sense of place, their heritage.

Shell doesn't have time for conversations about sacred land. *Next thing you know, some farmer will claim that leprechauns live among their sheep! Dear little icons that must not be disturbed by progress.*

MEETINGS OF THE PRO-GAS GROUP "One Voice for Erris" are

held in the Kilcommon parish church hall, but when the Corduffs attempt to establish a rational dialogue about the proposed refinery, their questions go unanswered. They're told to quit complaining, get with the program. Whether they like it or not, gas is coming to Mayo. Nothing anyone can do about that. Everyone should support this great economic opportunity.

Tossing and turning through sleepless nights, Willie and Mary vow to stop attending these meetings. They want fossil-fuel spokespersons to stop circling around legitimate questions, to stop playing mind games. They wish to engage in honest conversations about their community, their nation, and their planet's future.

Willie and Mary are learning about the grubby world of politics. Didn't their parents, their schoolteachers, and their Church tell them they must not lie? Tell the truth, accept the consequences, live an honest life of integrity. Why would anyone choose to do otherwise?

This much they know: If the company prevails, the community's way of life will be destroyed. They have no choice but to fight.

"There was an old lady back the road here," says Willie. "We used to take care of her, and the day after the parish priest and the bishop flew out to bless the gas site, she says, 'Oh William dear, I think we'll have to sign for the gas.'

"'Why?' I ask her.

"'Oh, did you not hear the priest?'

"To older people the priest is God. If you want to talk to God you must go through the priest. Later, people will say that the priest who supported Shell so much realized that he'd made a mistake. Well, maybe he did know that, but he never came back to the church to say anything about it. He didn't come back to the parish and admit that he'd done the wrong thing.

"My parents were devout Catholics, attending every mass Fridays and Holy Saturdays. They would have died for it. My father went around with a bicycle collecting money for the Church. Christmas, Easter, and a harvest collection. He kept a little book and he'd go out rain, hail, whatever was there."

Willie helped his father collect turf for heating the priest's home. Later, during the uprising against Shell, an article appeared in the local paper implying that Willie was acting like his own father who, claimed the writer, had rebelled against the electric company.

"He did not rebel. When the company came in wanting to put poles on the land, he needed to know how much it would cost. They told him it would be the same as people in the village. But then they said he'd be charged extra, and he couldn't afford it, so we didn't get the electricity for one-and-a-half years after the rest of the village.

"People who should have known better portrayed my father as a protester, and so now here we are, a family of protesters. The priest who wrote that article didn't talk about how my father had gathered the collection for him all those years."

"I think they got so excited with the clergy," says Mary. "Shell had a hundred years of experience. And if you have experience and money, you will beat anyone. Shell's PR machine was writing gospel about the company, and it was all so good."

BEFORE SHELL ARRIVED, NEIGHBORS respected one another's property, never intruding on someone's land without permission. Now, a corporation was demanding the right to walk upon their land, measuring, digging holes, plotting the route for a dangerous pipeline that would wind through the community on its way to an experimental refinery.

Shell sent letters to property owners, showing the pipeline's route, how deep it would be buried, and the number of kilometers it would need to travel from where it came on land to the Ballinaboy refinery. These letters included the amount the company would pay property owners per linear meter.

"Crumbs," says Mary. "Just crumbs."

A little extra for early sign-up, and in the same letter the threat of *compulsory acquisition if someone dared not to sign.*

"People who thought they were going to get a great deal of money were mistaken," Willie recalls. "They were selling their father's land, their grandfather's land, for nothing. Giving away their rights as Irish citizens to own land without a struggle. Shell was setting a precedent. If they could take private property in Mayo, they could take it anywhere in Ireland. No one would ever feel secure again on their own land.

"It was a *coup d'état* that no one believed could happen in Ireland, a crime committed by private industry, supported by church and state against the Irish people."

"We could not understand why people in our village didn't join us," laments Mary. "Willie's family has been here for four or five generations. You met neighbors you'd been helping for years, and to think you were going among those people and they weren't going to speak to you."

"It has made life much easier for me," laughs Willie, "because I don't have to get out of bed at four o'clock to help people who said, 'Look, Willie, we're totally behind you.' People who didn't want the gas, but did want the money. I said, 'I'd rather see you in front of me.' They knew that I didn't have money, that I needed money, but I wouldn't accept money. People thought I was daft.

"Shell was brought in here by the government, [which] changed every law to suit Shell. We're the most corrupt country in the world. [The people] have no say. They tried to put us out of our home, and I told them, 'The only way you'll get me out of here is in a box.'

"That was our biggest problem. We had nobody to turn to. If somebody came into our home, attacked our family, we had nowhere to go. I couldn't pick up the phone and call the police. They wouldn't respond.

"I begged the police for help. They knew me. I've been here all of my life. They knew I wasn't there to get them hurt. Every protest we went to, in the morning I begged for help.

"I could see the thing was getting worse and I often went to them on the road, just begging them to call a meeting with the government, with us, and with Shell. They told me to fuck off.

"I have known at the very least fifty of those police out on the roads. Twenty-five of them have known me since I was a young lad growing up. I had known them by their names. One guy I knew very well because he liked horses, he traded horses, and I always had horses.

"I pleaded with them. This thing was going down the road like the North. I was so dumbfounded. I said, *Jesus, is this real?* I thought they would change, that they would come over to our side. They'd *have* to.

"Sure, we thought local Gardai wouldn't do us any harm. I would wake up thinking, *All right, they'll be helping us today,* that they would tell the new ones not to bother us because we'd never done anything wrong."

"It's like the local guards here fed information to the new ones coming in," says Mary. "So, you felt, *Hold on here, did they have a meeting that morning? Did the local guards tell them who they should look out for, who to attack?* It was all like a bad dream."

Shell informs Willie that its workers will be coming onto his land. Not going to happen, he says. Shell's lawyers secure an injunction preventing specific individuals and all others from interfering with the company's work. Men turn up without notice, accompanied by police who order Willie and other farmers to give way. Landowners demand to see papers granting Shell the legal right to enter their property.

Don't need that, scoff the intruders.

Willie says he's willing to go to jail.

Fine, says Shell. *See how he feels once he gets thrown in with real, hardcore criminals. A few days away from friends and family will change his tune.*

On a January morning in 2005, the managing director of Shell appears on Midwest Radio, talking about the company's desire to have a dialogue with property owners who are refusing to respect the injunction. That very afternoon, accompanied by the police, men working for Shell walk onto Willie's fields.

WILLIE CORDUFF AND FOUR FRIENDS stand before Joseph

Finnegan, a High Court justice. Apologize to the court for violating the compulsory acquisition order, says Finnegan, or they will be fined hundreds of thousands of euros. He will take their cars, he will take their homes, he will take the land of every farmer in County Mayo.

Willie and his friends refuse to acquiesce, or to apologize for protecting their own land. Mary refuses to believe the state would jail her husband. "No," says Willie, "we are going to jail," even though in private he keeps telling himself that it won't happen. In one confidential memo, Shell debates whether it ought to jail all of the resisters, but the company decides that might be bad publicity. Four farmers and one schoolteacher will suffice to quiet things in Mayo.

Squeezed into a dark police van, each man placed inside of his own wire cage, like a "confessional box," Willie fears the unknown. *Will he and his friends stay together? Will they remain strong?* He is going to miss Mary and the children, working the farm, his friends, the family dogs. Why apologize to the court? He has done nothing wrong. Shell Oil sent him to jail; he will stay there until the company decides to let him go home.

Mary phones to tell the men that they are receiving a great amount of support. It's 2005. The country is beginning to learn about Shell's bad behavior in a small community, about police violence, about how the government is facilitating the abuse of Irish citizens.

People are learning why five men are willing to remain in jail for an indefinite period of time. Local protests are growing. There are pickets at Shell's worksite, and the company suspends all work there.

Mary pours more tea and coffee. "The men knew," she says, "they'd done the right thing."

Warders awaken inmates around six a.m. The men have a television in their cell. They can make tea. At nine o'clock they line up for their cereal. At ten they are allowed into the yard, where they meet inmates, many addicted to drugs and alcohol. Young men who need job training. Keeping them in prison will ruin them.

"But why are *you* men in here at all?" ask the other inmates of Willie and his fellows. The others assume that these old ones must have been in a bloody pub row. Surely no one in Ireland goes to prison for trying to protect his own property. Watching protest marches on television, listening to public speakers who support Willie and friends, the inmates bang on and kick their cell doors to show their respect for the resisters.

Willie and his four friends—the media call them "the Rossport 5"—follow the rules, lining up for meals, not allowed money or watches, given but six minutes of phone conversation each day.

From Rossport to Dublin is an exhausting five-hour drive, requiring that Mary start out at four, the latest five, in the morning. Willie lies awake at night, thinking about all the accidents on the roads, worrying that his wife might be hurt in a crash. Is it worth it?

Mary and Willie are separated by glass when they talk, no physical contact allowed. Crowds of visitors, much noise, but still they can hear other prisoners waving and shouting, "Fair play to you, Willie Corduff!"

Stay strong, urges Mary. *People are waking up. Shell has lost the publicity battle. The world is on the resisters' side.*

DURING VISITS, MARY IS GIVEN a number, then she must wait to be taken into the prison yard and through three sets of locked doors. Years later, she still hears the dreaded sound of keys locking and unlocking gates and doors; she still feels the helplessness she felt then.

Never a public person or politician, Mary always wondered if she was doing the right thing after appearing on a television or radio show, or speaking with a print journalist. She slept little, and was always on the road, suffering from constant tension. Letters and cards arrived from throughout the world. Neighbors brought over food. Days, weeks passed.

Shell's spokespersons popped up everywhere, delivering their feeble explanations as to why the company could not lift the injunction.

After ninety-four days, the men walked out to tumultuous celebrations on the streets of Dublin. Bonfires burned along roads back to Mayo, welcoming the Rossport 5 home.

TWO YEARS LATER, WILLIE AND MARY travel six thousand miles to California, where he is awarded the prestigious 2007 Goldman Environmental Prize in a ceremony at the San Francisco Opera House. Established in 1990 by San Francisco civil leader and philanthropist Richard N. Goldman and his wife, Rhonda H. Goldman, the prize has been awarded to 119 people from 70 countries—each year, one from each of the world's six inhabited continents.

Reacting to the news that he'd won the Goldman Prize, Willie told the press:

> We didn't start the campaign to win any prizes, but it shows that someone out there could see that we were doing the right thing. We always knew we were, but this means people elsewhere in the world saw it that way, too. . . . Seven years is a long time to be fighting something, trying to get people to listen to you. I hope more people in Ireland will become aware before it's too late, before the damage is done. . . .
>
> We'll have more power after this prize; more people in the world will realize what Shell is doing to our community. I hope more people will take on what the Irish government haven't had the courage to.[1]

Welcoming a crowd of 3,000 people, prize founder Richard Goldman says that recipients "have succeeded in combating some of the most important environmental challenges we face today. Their commitment in the face of great personal risk inspires us all to think more critically about what ordinary people can do to make a difference."[2]

Willie is in the company of environmentalists from around the world who risk their lives on behalf of their natural and human communities. A few years after his acceptance speech, in early March 2016, Goldman Prize–winner Berta Cáceres would be murdered in Honduras, after years of working to oppose the Agua Zarca Dam, a project that would, as her biography at the Goldman Prize site puts it, "cut off the supply of water, food and medicine for hundreds of Lanca people and violate their right to sustainably manage and live off their land."[3]

Just months later, Goldman Prize–winner Isidro Baldenegro López, an indigenous Mexican activist who led the fight against illegal logging in old-growth forests, in a region plagued by violence, drug trafficking, and corruption, is murdered. (His father had also been murdered for the same activities.)

In 2015 at least 122 activists are murdered in Latin America while trying to protect natural resources from environmentally destructive mega-projects such as dams, mines, logging, hydropower projects, agribusinesses, tourist resorts, and other development, according to research by the international nongovernmental organization Global Witness. Worldwide, at least 185 activists were killed that same year.[4]

ON APRIL 22, 2009, WILLIE CORDUFF and his neighbor Gerry Bourke crawl under a Shell-contracted delivery truck to protest continuing work on the pipeline and refinery. Men wearing dark clothing and balaclavas beat Corduff with batons, kneel on his ear, and yank his hands behind his back. The attackers are highly trained in the art of inflicting pain without leaving obvious wounds on the human body (the bruises will show up later).

Willie can't breathe. To prevent him from dying, the assault team occasionally relents. He stops struggling and they leave him, not knowing whether he might die.

WILLIE LIFTS HIS TROUSER LEG TO show the scar where one of the attackers slammed a large stone against his ankle, trying, it would appear, to smash bone.

"I think that if Shell had known at the beginning that we weren't after money," he tells us, "they would have taken us out, killed us. After the jailing in 2005 they couldn't very well kill us, but had they known earlier what they found out, they would have killed us for sure in the first year.

"I couldn't tell our [entire] story in six months," he muses. "It would take that long. We'll never be the same. It changed us completely. We always helped people, my father and I. To see a lot of those people turn against you is hurtful."

Mary adds, "The majority of the Church was on Shell's side. [There was] very little support on our side. Is that what your leader is supposed to do, go out to help an organization like Shell? That's very sad to say. They [local and national Church leaders] had to know about Shell. The Catholic religion is not about money. At least that's what we were taught, but we found out different.

"Without the full cooperation from the state and Church, Shell couldn't have accomplished what they did. They got the full compensation; we got none."

Did Shell pay the clergy? I ask.

"Oh God yes," Mary laughs.

Mary doesn't think the wounds from the uprising have begun to heal. Perhaps that will happen one day, but no one knows when.

Pat O'Donnell in Broadhaven Bay, County Mayo

Fisherman

> When I undertook to confront Shell and the Nigerian establishment, I signed my death warrant, so to speak.
> —Ken Saro-Wiwa, Nigerian writer, social activist, Nobel Peace Prize nominee, and president of the Movement for the Survival of the Ogoni People, in a letter to Sister Majella McCarron, summer 1994 (undated)

PAT O'DONNELL STANDS AT THE HELM of his fishing boat off the coast of County Mayo. A massive five-story freighter looms close by. Royal Dutch Shell uses this ship to lay gas pipelines below the seabed.

"Stay clear of the *Solitaire*," calls a deep foreboding voice. "Warning. Stay clear of the *Solitaire*."

Pat's boat is a seabird, taunting a great white predator.

"I have crab pots in this area," he shouts. "This is where I earn my livelihood. I will not be leaving here."

No sign of a captain or crew on the freighter.

PAT O'DONNELL IS A NATIONAL HERO, recipient of a letter of

thanks from Ireland's first marine rescue awards, *"for the strength of spirit that makes our communities better places in which to live, that selfless sacrifice and endeavor that lifts us all and benefits us all."*

On the morning of October 25, 1997, a German businessman had taken a couple and their eleven-year-old daughter out in a currach, a low-slung boat used by generations of Irish fishermen, to explore the fascinating sea caves along Horse Island. Caught in a swell, their boat capsized, stranding the family in a large cave. Pat and his 12-year-old son Jonathan, hearing the emergency call, sailed out on the *Blath Bawn*, and Pat's brothers, Tony and Martin, went out on another boat to assist in the rescue.

Local police coordinated the rescue on land, while the Irish Coast Guard carried divers to the mouth of the cave. Divers Josie Barrett and Michael Heffernan died trying to swim into the cave with a line. Rescuers found the family huddling in a crevice just a few feet above the water.

The O'Donnell men worked through the night, risking their lives to help save the stranded family, and to rescue police who needed help, actions for which they were honored by the Coast Guard and later by the Irish government.

PAT'S FAMILY HAS LIVED IN COUNTY Mayo for three hundred years, generations of fishermen and farmers. He grew up hearing stories about the Great Hunger (the devastating years-long Irish "famine" that began in 1845, when the potato crop failed). More than one million Irish citizens starved, while two million more boarded "coffin ships," seeking safe haven from hunger and disease. There had actually been enough food to keep people from starving or having to flee, but wealthy landowners exported commodities like wheat, corn, peas, onions, rabbits, and salmon to England, while the bodies of Irish children, their mouths stained green from eating grass, rotted in ditches.

Pat's ancestors struggled, suffered, and survived this great disaster, but even in his own generation things remained economically bleak for his family. He remembers watching his five sisters leave for America, where they hoped to find work. They would send money home and, God willing, return to their beloved Mayo. Pat and his brothers could not afford to grieve for their sisters. Work, not self-pity, put food on the table. Their childhood was over. It was their turn to ply the treacherous waters off northwest Ireland, fishing to support their family.

Pat has spent his entire life on the sea. He understands the Atlantic's erratic moods, how unsuspecting people might sail out on a fine blue day only to get caught without warning in a deadly storm. Fishing is in his blood. He's built a business, exporting crabs and lobsters to Europe. His two sons are fishermen; he hopes their sons will fish as well.

BOARDING PAT O'DONNELL'S BOAT, the police mill about, seemingly reluctant to arrest a man they know and respect.

"Pat," they implore, "be reasonable now. You're breaking the law."

"What law?" he demands. "I've been working here all my life. Right here in these waters. You know that. Has it ever been against the law to make a living?"

The police hesitate.

"Tell Shell to go away, before they destroy our bay," Pat insists. "I will not move from here. Go ahead now. Do your duty. Arrest me. Put me in jail. I will not be moving, I tell you."

Police escort O'Donnell to shore, where a cheering crowd of supporters is waiting. The authorities confiscate Pat's boat, but he will soon be back in Broadhaven Bay, taking risks to protect Irish waters teeming with dolphins, seals, otters, whales, and basking sharks.

Broadhaven Bay is designated as a Special Area of Conservation (SPA) under the 1992 EU Habitats Directive, while Blacksod Bay/ Broadhaven Bay and Carrowmore Lake have also been recognized for their "high ornithological importance" and have been designated as SPAs under the EU Birds Directive (2009/147EC).[1]

Parts of Mayo are designated UNESCO conservation sites. There are soft beaches, sea stacks hundreds of millions of years old, many kinds of birds including puffins, peregrine falcons, and the great cormorant.

SCHOOLTEACHER, IRISH SPEAKER, singer, and community activist Micheál Ó Seighin invites Pat O'Donnell and other friends to his home, where he tells them that Shell is planning to dump toxic chemicals into Broadhaven Bay and Carrowmore Lake.

Micheál is known as a man of unquestionable integrity. Still, people find it difficult to grasp what he's telling them. Why would anyone want to poison the drinking water for ten thousand people or pollute pristine waters teeming with wildlife?

Everyone is upset. Does this company think people in rural Ireland are backward, that they do not know how to read, or that they will not protect the lands their ancestors tilled, the waters they fished, the heritage for which they fought, passing it down to future generations?

> Shell's core values are honesty, integrity and respect for people. The Shell General Business Principles, Code of Conduct, and Code of Ethics help everyone at Shell act in line with these values and comply with relevant laws and regulations.[2]

Concerned about the threat to Mayo's waters, Pat and fellow local fishermen discuss ways to resist Shell's plans. But solidarity breaks down when the company offers boat owners €100,000 (approximately $150,000 at the time) to stop working for three years. Pat O'Donnell could retire, build a new home, and spend winters in Spain with his wife and children. Pat's family refuses to take the money: "We do not tolerate the direct or indirect offer, payment, solicitation, or acceptance of bribes in any form."[3]

BY JUNE 2009, PAT O'DONNELL has spent five-and-a-half months in prison, has been badly beaten by the police, and has turned down substantial offers if he'll silence his opposition to Shell's project.

Pat hears that someone is planning to damage his fishing gear. Determined to protect his property, one morning he sails into Broadhaven Bay. Without warning, like an apparition, a craft appears alongside his boat the *Iona Isle*. Four figures wearing wetsuits and balaclavas, two of them holding guns, climb aboard.

With military precision they commence to sink his craft, but Pat pulls an inflatable raft—something the attackers overlooked—down from the boat's wheelhouse ceiling, and he and a friend escape just moments before their boat slides beneath black waters.

Water police arrive. Pat points out the way the attackers fled. The police speed off in the opposite direction.

Two days later, Shell E&P Ireland issues a statement emphatically denying that employees of the Corrib gas project had anything to do with sinking Pat O'Donnell's fishing vessel.

One month before the attack, Pat had refused to sign onto the agreement brokered by Shell offering local fishermen substantial amounts of money if they would agree not to fish in Broadhaven Bay for three years.

Two weeks after the sinking of Pat's boat, Shell writes to him, his brothers Martin and Tony, and his son Jonathan, asking them to remove their shellfish gear from the bay. The company's ship, the *Solitaire*, is en route to lay offshore pipe. Shell offers "fair and equitable" compensation for disrupting the O'Donnells' fishing activities. If the men refuse to cooperate, Shell will go ahead and remove all of their three thousand crab pots from the bay.

> The state is bound to protect the personal rights of the citizen, and in particular to defend the life, person, good name, and property rights of every citizen.[4]

TWENTY-FOUR HOURS AFTER RECEIVING Shell's solicitation, Pat and his son are arrested on two different boats in Broadhaven Bay. Detaining the O'Donnells' boats, the police beat Pat so badly that he is taken by ambulance to a hospital in Castlebar. Jonathan is driven to prison.

The next day police boats as well as craft carrying mercenaries accompany Shell's *Solitaire* as it sails into Broadhaven Bay.

"That was the problem they had with me," says Pat. "I did not accept their bribes, and I won't accept their bribes. If something wrong is going on, or something bad is happening, and you have been bribed, then you have to keep quiet about it. You have to be silent.

"So, I can speak. And if I see something wrong I can talk about it, or do something about it. We have four licenses. I stood to get 300,000 euros [$450,000], and my son would get 100,000 euros. And if I had gone in to talk with them, they'd have probably doubled whatever they offered before. But it was the wrong thing to do.

"We know how good the sea has been to us. And we just throw Shell and their money aside because my two sons are fishing now, and their sons may be fishing. And the main fear would be that when they go pumping in the raw gas, there's an explosion maybe a couple of miles offshore, and all the chemicals will be coming in with the gas. This would destroy the fishing industry, just like the way BP did off the Gulf of Mexico."

STRANGERS CRUISE ABOUT SMALL Mayo towns at night, filming people going into stores, restaurants, and pubs. *Who are these arrogant men? Why are they poking cameras at people who are going about their own business? Why won't the police put an end to this harassment?*

"I was five months in a high-security prison, [with] a lot of murderers, robbers, rapists, the whole shebang," Pat remembers. "The prisoners and 90 percent of the guards had respect for me. 'Pat, now,' said the guards, wanting to keep my spirits up, 'you should not be here. You're not a criminal. It's the bastard politicians that should be in here.' Some of the prisoners were very good to me, but I never asked for special favors.

"I was in jail in the spring of 2010, and they jailed me for that summer. Shell wanted me out of the way, and the state wanted me out of the way. It was a manufactured conviction, that's what it was.

No doubt about it, I was a political prisoner. I had done nothing wrong. The judge took the word of a policeman. No real evidence, no witnesses, just the word of one policeman, and I was sent off to jail.

"The charge? *'Violating public order. Threatening, abusive behavior,'* something like that. *'Invading the space of a policeman. Loitering at sea.'*

"They just make up these charges as they go along. You see, the way it works is that everything comes from the top, from the Minister of Justice all the way down to the bottom rank, who are told what to say. They get the convictions, and they get you off to prison."

AS RESISTANCE TO THE PIPELINE/REFINERY grew, Ireland's mainstream political parties, Fianna Fáil and Fine Gael, denounced Shell's opponents. Media accused Pat O'Donnell and friends of being members of the Irish Republican Army, determined to wreak havoc in their own community. None of these sources explained why people who were risking their lives to protect their families and friends would conspire, at the very same time, to harm them.

"People you live with, and the protesters, knew you're not any of this kind of stuff," explains Pat. "We were a community under siege from an oil company. And one day I was talking to two Frenchmen who'd seen Richie O'Donnell's—no relation to me—documentary, *The Pipe*, about the resistance to Shell in County Mayo. They couldn't understand why the rest of the Irish nation didn't get behind our community. They said the whole of France would become one in a fight like this one.

"We were left to fight Shell on our own, and they brought people up, they bribed them, and there were only a few of us left that didn't bow to their might.

"What happened here will come to light at some stage. It will be very late. It will come out that Shell cares only about making money and they don't give a fuck about anything but profit. It's all about profit with Shell, or any multinational."

THREE YEARS AFTER OUR FIRST conversation in 2013, a time when Shell's opponents still felt they might prevent the company from completing its refinery, Pat O'Donnell and I talk again in Kilcommon Hostel. He appears tired, perhaps from being asked to revisit painful events.

Is he angry or frustrated? Does he have nightmares or suffer from other symptoms of post-traumatic stress syndrome?

"Am I angry about it now, is it? No, I came out doing all right on that end. You see, I kind of understand the way things work. It's all about big business. The politicians and the government bend to big business. As far as I can see, the small man doesn't matter. The small fishing communities are lost around Ireland. They're dying because of what the government did with our fishing resources. We [the Irish government] signed up with the European Union and gave away most of our fishing quota. We have only four percent now. We gave away our Irish waters. They don't belong to us anymore.

"What happens over in Brussels in the European Parliament is a higher level of corruption than what we're used to in Dublin. People who refuse to understand this are surprised and shocked by the stuff I talk about."

We talk about Pat's being sent, twice, to prison.

"I had a solicitor, but the courts don't listen to defendants. I spent every fucking day of that sentence," he laughs mirthlessly, "except for the 25 percent remission for good behavior. As if I was a criminal.

"It's tough, not easy, to be in jail. I wouldn't want to be treated different than any other prisoner, wouldn't want special treatment. I'd never ask for forgiveness. The court offered me 240 hours of community service, but I wouldn't agree to that.

"You have to *buy* justice in this country. You need a lot of money."

Pat was arrested several times while fishing on his own boat, taken in and held for four to six hours.

"But," he says, "they were just buying time so Shell could get on with their work of laying the pipe beneath Broadhaven Bay. They couldn't charge me because I was just doing my own work.

"I was talking to a prison officer, we were about the same age, and he was set to retire. His job for many years had been bringing prisoners from the jail to the court, so I asked him about what it'd been like in the old days.

"'Well,' he said, 'when you brought a prisoner to court in the old days the parish priest decided who went to jail and who went home. But today,' he said, 'Money decides who goes to jail and who goes home.'"

Pat doesn't want to talk again about the sinking of his boat, so I ask if he's any closer now to finding out who the pirates might have been.

"No." Pat laughs, as if I'd asked whether he still believes in the Easter Bunny. "To this day we don't know who did it."

"Any idea?"

Pat answers with a big "stop-asking-silly-questions" grin.

"We didn't know the police who broke my teeth, put three stitches on my brother's eye, and broke a bone in his neck," he says. "But we found out that one of the young cops was a champion boxer. We took a case against them, and it went all the way to the High Court in Dublin, but the judge made up a cock-and-bull story that one of the cops was being threatened, that we were trying to kill a police officer and the Gardai had to assault us in order to rescue their colleague from being killed.

"When he was instructing the jury, the judge told them that everyone knew that my brother and I had never been assaulted. This case resulted in a hung jury.

"I got fed up because the police lie, and their word is always accepted before an ordinary citizen's. Some big multinational company can buy the judge, buy justice. They don't care about anything. My solicitor showed them my commendation for bravery, pulling people out of the sea. I saved a police officer! I'm not a bad person. And the judge knew that, but he had to do what Shell wanted him to do. It's not what Shell says directly to the judge. It's what Shell says to the Minister of Justice."

"THE IRISH PEOPLE DON'T UNDERSTAND any of this," Pat continues, shaking his head. "I'll give you an example: I was convicted of the charges, but the judge didn't send me right away to jail. He wanted me up to court for sentencing the next day. As soon as I went up, the first person I saw was Jim Fahey, the local correspondent for IRT television. And he said, 'Oh, it'll be just a slap on the wrist, Pat. That's the way the law works.' And I said, 'Oh no, this is not the law. This is Shell's law.'

"Then we went into court and the judge called me a thug and a bully, and that was broadcast out all over Ireland on the six o'clock news. That was one up for Shell. The media were very biased. Some of the journalists were on our side, but they had a hard time. And Shell bought up journalists from the local papers.

"You know, people talk about 'compensation,' but I call it 'bribes.' When you sign the dotted line, you're finished. I'm not changing my mind about that."

I ask Pat about video footage I've watched, in which Shell's opponents do not back down when police and mercenaries attack them. I ask if they'd studied nonviolence.

"No," he says, "we did not study Gandhi. We accepted our beatings. We took our beatings. We were sad to see the old people and women take beatings. We did not expect that kind of thing when we started out. I had some idea what the police were like, but I did not think *local* cops would do such things. It's very sad because if I were to meet them now, I wouldn't sit down and talk with them. I'd have no respect whatsoever. They could have taken a back seat and let those outside thugs in uniform come in and do the beating for them."

How does Ireland differ from the country in which he grew up?

"I was thirty-five or forty before I began to understand what was happening, but then you think about other places to live, and you realize that the same thing is going on all over the world."

"And yourself, Pat," I ask, "have you ever thought about running for office?"

"Oh no, no, Jesus no."

And the story that he'd been let out of prison for eight hours to attend his youngest daughter's first communion?

"Oh yeah," he says. "It was an awful job getting to do that. Someone said, I think it was Martin Ferris, that it was easier to get the IRA men out of prison than to get me out of jail for a brief time."

Would he do it all over again?

"I would," he says. "You'd have to. [For] the five or five-and-a-half months I spent in jail, my wife and kids supported me, and my brothers and sisters as well. I showed them I was all right when they came to visit me. I made friends in there. I had no problem with the prisoners. It was an experience to see the other side, what life is like there.

"Some of the lifers, after fifteen or sixteen years, would be coming out, and I'm waiting to catch up because I kept in contact with them, visited them in an open prison near Belfast. You know, they were just kids, and they want to get out. Play by the rules, educate yourself, and you might leave after fifteen years or so."

WE TALK ABOUT WHY PEOPLE follow orders, even when it means hurting someone you might know and even respect.

"One morning, I ran into a distant relation over in Ballina," he tells me. "I used to hang out with this guy in the old days before he became a Garda.

"'Jimmy,' I say to him, 'you should have retired with your integrity still intact. You lost it back at the protests.'

"'Oh no, Pat, that's over and done with,' he insists.

"'No Jimmy, it is not,' I say. 'It's a shame for you. But answer me one thing straight: If the state put a gun in your hand and they gave you the order to use it, would you have?'

"And he looks at me and says, 'Pat, I'd have to.'

"Well, they're all the same all over the world. In America or

Ireland, it doesn't matter. When the order comes, they will obey. If they're told to attack you, they'll do that. If ordered to shoot you, they will shoot you."

Shell to Sea activists next to crosses of Saro-Wiwa and friends executed by the Nigerian government

Witness

> And I should like to be able to love my country
> and still love justice.
> —Albert Camus, *Resistance, Rebellion, and Death*

TERENCE CONWAY GREW UP ON a small farm in Erris, County Mayo. One of fifteen children, he remembers having a good life: People were less materialistic. Children played outside in the villages, rather than being stuck inside with gadgets like iPads, cell phones, and television. His family had few material possessions, but they ate well and enjoyed lots of friends.

Terence attended primary school and worked in a factory before moving to Dublin, where he was employed as a laborer and a carpenter. In the late '80s work was scarce in Ireland, so he moved to the States where he got a job with the New York Housing Authority.

Terence survived fourteen years in the Big Apple at the height of the crack epidemic and the homeless crisis. Parts of the city were war zones—gunfire, wounded and dead on the streets. Shelters were overflowing, and homeless were living in tunnels beneath Grand Central Station, squatting in the Port Authority bus station, and turning the Lower East Side's Tompkins Square Park into a squatters' encampment.

Each summer Terence would spend a month on holiday in Mayo. When, after so many years away, he returned for good to help care for one of his brothers, he found that Ireland had changed a great deal. People seemed to be chasing after the things they had seen on television.

"In the old days," he says, "families were big and their houses small. Now, families are small and houses are massive. One or two children living in six-bedroom houses! The extra rooms must be kept for spiders and flies."

TERENCE IS WORKING OUTSIDE County Mayo when Shell flies a priest and a bishop out to bless the company's gas site in the North Atlantic. This bishop is Thomas Finnegan, who also played a prominent role in founding the pro-gas group One Voice for Erris.

"I wasn't here when they went out," says Terence, "but in my mind it's fine for the Church to bless the rig because you wouldn't want to see people get hurt on it. I suspect there may have been some hands greased. It was the Church's enthusiasm that makes me think that some money might have changed hands, but I don't know for sure. That's just my suspicion."

IN THE EARLY DAYS OF SHELL'S project, Terence isn't overly concerned that a gas refinery might be built just a few miles up the road from his house. He does not attend the One View for Erris group's meetings. Nor does he believe this fossil fuel company is going to drop bags of money on Mayo's doorstep.

"You just have to take a casual look at Shell's operations over the years worldwide," he says. "When they come into an area, most people suffer, very few benefit. Apart from a number of contractors, there'd be no money for people like me or anybody else from this project.

"Shell's only concern is that they themselves will make a lot of money. In reality, the community is irrelevant to companies like that. Short-term businesses like hotels, B&Bs, bars, and restaurants might stand to make a bit of money during the construction phase, but no one else will benefit.

"A few people did put up big houses, hoping to make a fortune but they couldn't keep up with the payments and the government took them over."

TERENCE BEGINS HIS OWN RESEARCH into Shell's project, scanning newspapers, listening to radio programs, and watching televised arguments for and against the refinery. He attempts to talk with politicians, and the Mayo County Council, but he's unable to get clear, unambiguous answers to his questions.

"So gradually I concluded that Shell had everything sewed up. The offices of the state were fully on Shell's side. Not partially, not a little bit, but completely. That became quite clear."

Shell is planning to build the refinery in a catchment area of drinking water for 10,000 people, on 400 acres that the company acquired from Ireland's National Forestry Association.

"This sale is done behind closed doors," says Terence. "No one consulted with the community, and we don't know how much Shell paid for this land.

"We know that location is all wrong. The top layer of the building site is peat bog, Underneath that is an unstable soil we call *doib*, and this contains a lot of aluminum, a neurotoxin if it gets into drinking water.

"Aluminum does wash into our lake. A monitoring committee is set up for the project. You have the Mayo County Council, the Environmental Protection Agency, and An Bord Pleanála. At first, the council threatens Shell with legal action, but then they do nothing.

"Testing the drinking water supply shows the amount of aluminum is rising, but the council goes from threatening action to covering it up.

"We learn fairly quickly that none of the mainstream political parties are going to help us. Even the Green Party is fairly useless. Crookedness of politicians seems to be almost universal. Only if you'd invite a foreign invader in to plunder our resources could things be any worse in Ireland.

"There is a long tradition in Ireland of fighting oppression, but now people want the fancy car, the big house, things like that. And as long as they can maintain that lifestyle they seem to be happy. It's about what I can grab, materialism rather than caring about the big picture.

"And then, of course, anybody that chooses to fight the corruption will find that there are many ways of punishing you. Your job might be in jeopardy. You may not get things that you'd normally get. There are clever ways of controlling at least some of the population.

"In the case of what Shell was doing here, if you did complain and try to do something about it, the local politicians might not want to associate with you. I witnessed this myself in Mayo, when politicians failed to stand up at rallies for the jailed Rossport 5.

"We picketed petrol stations all over the country, but the politicians didn't join us there. After the Rossport 5 were locked up, there were protests in the local towns and nationally. And I remember this one official saying quite clearly that if necessary he'd be willing to go to jail to stop what was happening here. But then very shortly afterward, there wasn't a word out of him.

"And there was an election coming up, so Michael Ring was out canvassing one day and he came to my door. I went out to speak with him, and he was asking would I give him my vote.

"And I said, 'Now, Michael, why did you go silent after that big meeting at Castlebar? Did the head boys and girls in Finn Gael pull you aside and tell you what's really supposed to go on?'

"He took off running," laughs Terence. "And I hadn't even raised my voice to the man."

In a November 2006 copy of the *Gardai Review Magazine,* a private publication, resisters discover that the police have established a policy of not arresting Shell's opponents. Assaulting people appears to be their only alternative. By not arresting people, they will avoid making martyrs out of opponents.

"It's a private publication, so they don't think we can see it but we do, and it is clear what we are dealing with, and what we will be dealing with: *political policing*. Shell and our national police are acting as partners in crime, helping one another to break Irish law. I recorded several instances of this on video.

"When it comes to Shell, the law got torn up and thrown away."

TERENCE ACQUIRES A VIDEO CAMERA and, determined to show the havoc Shell is wreaking in Mayo, he soon becomes a talented videographer.

"Before all this, I never owned a camera, but I knew that it was important to document what's going on. So, I get a video camera, and the police are very concerned about my showing up to film protests.

"At one protest, a guard standing next to me says. 'I see now, Mr. Conway, that you are very fond of YouTube.' 'Well now, do you think so?' I say. I didn't even know what YouTube might be, never having spent any time on a computer. I used to hate cameras, and the people who were using them.

"I was shocked by how far the guards were willing to go." (The guards, An Garda Síochána in Irish, are the Irish state police, also known as the Gardai. An individual guard is also known as a Garda.) "I knew they were capable of violence, but not to that extent. They did not care about my videotaping them because they knew the level of corruption that would protect them.

"The cops are worried that the truth will get out. But top officials tell them not to worry. Do whatever you want. Nothing is going to be done about Shell or the police.

"I make four videos and send at least 15,000 DVDs out around the country, highlighting exactly what is happening and what the important issues are. I send these DVDs to hundreds of local and national radio and television stations, and to newspapers. I tell them that I have additional information to send on if they should want it. Only one reporter ever gets back to me. And no one asks for more information.

"It turns out that Shell is passing out booze to media people throughout the country. I offer reliable information; Shell gives out bribes.

"Crime reporters in particular oppose the resistance movement. I send the DVD to many crime reporters, but they never get back to me. And they don't bother to tell the truth in their stories. It's fiction a lot of them should be writing."

One legal firm offers its services to the resisters, and Terence is prepared to make a case against Shell and the state. He passes evidence to the firm, and then doesn't hear back from them. Most likely, he concludes, this is just another effort to distract resisters.

TERENCE CARRIES HIS VIDEO CAMERA to demonstrations, recording scenes of violence against nonviolent protesters. One afternoon, a giant digger is dredging muck from the bottom of Broadhaven Bay to facilitate the laying of Shell's gas pipeline. Resisters paddling canoes and kayaks, some actually swimming close by, surround the digger. Without warning, the operator drops a load of rock and sand directly, it appears, upon these people.

Terence, filming this dangerous scene, shouts for viewers to hear. "He didn't get them! He missed. He didn't get them."

That Shell is willing to do something like this in full view of video cameras worries Terence. *Who sits at the controls of that digger? Who orders this person to drop a huge load of sand and rocks quite close to peaceful swimmers and people paddling small boats? If presented to a court of law, would this video footage not be grounds for a charge of attempted murder? Why would anyone follow orders to terrorize, injure, and possibly kill innocent people?*

The resisters do not flee.

Shell's public relations materials express the company's sincere desire to interact in good faith with communities in which it does business, yet Terence never receives a cordial phone call or letter suggesting he meet with company officials to discuss his concerns.

Shell holds what it calls "open days" in Belmullet, where its employees show large pop-up displays of the proposed refinery and agree to talk with local people; however, Shell's experts cannot or will not answer the most basic questions. Instead, they ask for attendees' names and addresses, promising to get back to them at some future date.

"They'd give you a cup of tea and a biscuit," Terence recalls, "and you'd walk around and look at their displays, but you would get no answers. Look beyond the gloss and it's not a pretty picture because, you realize, they do not have any real answers. And even if they did, they are not going to give them to you.

"I believe that directly or indirectly, some serious money was paid out to politicians. There was *no benefit* in this refinery for Ireland. Shell could take the gas 100 percent and write off every development cost. They were invited here to plunder our resources as they wished."

TERENCE DOESN'T THINK IT WOULD be possible to hold Shell accountable for what they did in Ireland without initiating a court case that would take many years to resolve. Whether it's in Ireland, Nigeria, or his previous home in the United States, Shell will only do whatever the state allows them to.

"The Irish government might not be the most corrupt in the world, but since I came back here to live it seems like they are competing for this title. Our government might indeed get that title before too long. Shell has helped people see just how rotten this government is."

In retrospect, Terence isn't sure what resisters might have done differently during the fifteen-year-long resistance campaign.

"This was a learning experience," he says. "We had to learn about fossil fuel companies. I'd like to see people look beyond the gloss, to see what the government is really doing. And once you do that, you'll realize that it's all about taking care of the rich, the very rich.

"I would hope that there will be a total change in people's thinking. Look to see what's behind the government's policies. It has nothing to do with benefiting the people. Just look at the issue of homelessness—family homelessness has gone through the roof over the years. Vulture firms are buying up property all over the country and throwing people into the street, hiking up rents that people can't survive.

"Scrap the present gang of politicians, and vote in politicians who will actually do what they say and work for the people, not multinational corporations.

"Ireland absolutely needs a revolution."

TERENCE HAS BEEN DRIVING ME the length and breadth of Ireland while I suffer from an abscessed tooth. I just want to swallow painkillers and curl under the covers, but he continues to contact people whom I should meet.

We travel in a small car. We drive for hours, sometimes all day, on narrow twisty roads and well-maintained highways. Terence smokes, we talk, we stop for tea. Great trucks roar toward us; I read that an American family newly arrived at Shannon was killed in a terrible accident. Terence has driven these roads all his life. He does not make mistakes. There are few lights in the countryside at night other than dim lights in little homes set back from the road. We drive for miles without seeing another car. I expect to see deer or fox in the road.

Terence drops me off at Kilcommon Lodge Holiday Hostel. The common room is quiet. Hostel owner Betty's little puppy must be sleeping. I tell my tooth to stop drilling holes in my face. It doesn't listen. My bed is comfortable and warm. Somewhere near my window Betty's two pigs are snuggled close together. In the morning they will appear again, rooting, chewing apples she tosses to them. Sometimes they stand nose-to-nose, like lovers or perhaps good friends. One day they will appear on Betty's breakfast table, a fry of rashers, beans, and eggs. I try not to feel sorry for them. They do not know their fate.

The Rossport 5: (l. to r.) Brendan Philbin, Philip McGrath, Willie Corduff, Vincent McGrath, Micheál Ó Seighin

Scholar

> It's not the most distressful country after all when five "green men" stand their ground.
> —Seamus Heaney, Irish playwright, poet, and translator; recipient of the 1995 Nobel Prize in Literature, on the Rossport 5

MICHEÁL Ó SEIGHIN IS A RETIRED schoolteacher, a singer, and a scholar who has been involved all his life in drama, music, and community development. He is a consummate storyteller who spent ninety-four days in prison for challenging Shell Oil's right to act as an eighteenth-century land baron in County Mayo.

If one day you visit County Mayo, hoping to meet people who fought long and hard against Shell, most everyone will direct you to "Shine" (an anglicization of his Irish surname). Known as a reliable narrator, Micheál does not embellish or exaggerate the fight against Shell. There's no need. During fifteen years on the front lines of the uprising, Micheál came to realize that the system he'd believed in all his life is corrupt, defunct, not worthy of anyone's allegiance.

Micheál grew up in Limerick, the youngest of nine children, reading a great deal while his siblings were away at boarding school. Two brothers became priests, and a third also wanted to be a priest but his health gave out. One of his sisters chose to be a nun, and the children of family friends also became priests and nuns.

"This religious thing must have influenced me," he says. "My father would have been more devout. Not sure my mother believed in anything at all. In politics, yes, but she kept to herself."

Micheál's parents were married on the day the Irish Civil War broke out. They walked home through fields past burned bridges to their small, not more than a hut, dwelling.

Micheál's mother had only one maternal first cousin, and this man had become a captain in the Free State Army. One day this cousin arrived with a company of soldiers to search the Ó Seighins' abode. Never again did Micheál's mother speak to this man, nor to his family. She refused to mention them, and she did not attend their funerals.

When Micheál first arrived in Mayo in 1962 as a recent college graduate, schools were having great difficulty getting teachers to stay. A systematic effort to develop the Irish economy was being launched, which meant there were many jobs for college graduates. He agreed to work at the local school for two years, teaching English, math, Irish history, geography, and the Irish language.

"And here I am still. A learner. That's what I do. I love learning."

Micheál never wanted to get involved in party politics.

"I am a Republican, became a member of Sinn Fein very late, after the ceasefire of 1994 in the north. I knew that the war was over then. It was obvious."

Micheál did not share his political beliefs with students, but one morning he arrived at school to find every car plastered with posters of a man running for some high office. Shocked to see that teachers were engaging in such blatant party-pushing in school, he acquired a poster showing the popular politician's opponent, and stuck it upon his car.

"Everyone knew," he laughs, "I would never be supporting someone like the man on that poster."

WE ARE TALKING BESIDE A WARM FIRE in Micheál's cozy living room. The phone rings, he stops for a moment to greet visitors in the kitchen or to tell someone goodbye.

"I learned from my mother a certain way of dealing with things," he says, returning to the fire. "Perhaps a rather extreme way. When things are out of control, you just swat them and get on with the next thing. If you cannot deal with something, you just swat it. That probably is the way I have been ever since childhood."

Micheál grew up listening to stories about the Irish Civil War (1922–1923). He remembers finding bullets in a ditch near his home, two crosses where local men had been shot dead by the army, and bullet holes in the gable of another house.

"Our landlord at the time didn't seem to understand that the world had changed, or that things were changing fast in Ireland. He'd show up with his dogs and horses, intending to spend a leisure day hunting fox. But the last time he brought his dogs in for a hunt, a local man killed two sheep, poisoned them, and put them out on a hill.

"The landlord left for home. All of his dogs were dead. That was the last time he ever came around.

"I was always involved in local development, county development, community centers, drama, singing, teaching music to kids. Going to meetings and that kind of nonsense. I'm an activist by nature, but I was pretty much tired of it by the time the Corrib gas issue came along. I had put in a lot of heavy lifting and wasn't prepared to get involved anymore."

In early April 2000, the parish priest announced in his weekly newsletter that gas was coming in to Mayo.

"The Council for the West was a Catholic Bishops' group with the backing of the Anglican Bishops," remembers Micheál, "and they were thrilled about the gas.

"I didn't pay any attention to that at all. I didn't want to be involved. I had no interest. Then I got word from a man who worked for the Department of Fisheries, asking that I tell the local fishermen that the government was doing a survey regarding the Corrib gas project, and unless they spoke up they would not be considered at all."

Micheál wrote a letter to the local fishermen and made sure it was hand-delivered so no one could deny they'd received it.

"As far as I knew, the gas would be treated out at sea, and then it could be brought ashore to odorize it so that it would be apparent if there were a leak in the pipe.

"Enterprise Oil was handling things at this stage, but I might as well say Shell, because they were mostly involved after that. There were meetings here and there, and I didn't get involved at all."

Micheál learned that company engineers were planning to bring the gas pipe around Glengad Hill, where there had been a landslide not that long ago.

"If the oil company sinks a gas pipeline through that hill," he told students, "they will bring it straight down."

Micheál, a student of geography and geology, knew a great deal about County Mayo's landscape. Nevertheless, he asked himself, "What do I know about gas pipes? Enterprise-Shell have the best engineers in the world. Surely, they know something that I don't about all this.

"I was wrong, because the best in the world does not mean a thing. It's the motivation that's important.

"At the same time, a German friend, Gerard Muller, kept tormenting me, demanding that I do something about Shell's plans. The company had applied for planning permission, and the books of supporting literature were in the Gardai barracks in Belmullet. 'Go and you have to do something,' he said. 'Go and have a look.'

"So, I said, 'All right. All right, Gerard, I'll go in.'"

Accompanied by his daughter Brid and two close neighbors, Micháel visited the Garda barracks in Belmullet to examine Shell's Environmental Impact Statement (EIS). A desultory task until he read the first few pages of this two-volume document, where he discovered a list of "the little uglies"—mercury, selenium, cobalt, radioactive elements—that Shell was planning to dump into Broadhaven Bay and Carrowmore Lake.

Not *accidental* leaks. Not *possible* mistakes, but deliberate, calculated plans to damage the health and welfare of County Mayo residents.

"The whole shebang of heavy metals. I was shocked. Horrified. A quick look and I saw what they intended to do. There would be runoff from the terminal site straight into Carrowmore Lake. Through *perforated* drains!

"Now, I know the dynamics of the Bay. There's a tow in, and not out. The poisons were going to stay right there.

"Some of us read and studied, then submitted our objections regarding the refinery, and just after Christmas 2002, the journalist Christy Loftus gave my challenge the front page in his newspaper."

Soon after its publication, Shell withdrew its first application. The following year they put in a second proposal.

"We always knew," continues Micheál, "there had been coordination from the word go between various authorities and the oil company. The second Environmental Impact Statement, now fifteen volumes, contained exactly the same information as the first two we'd read.

"You see, they said in their application that the toxic chemicals would be one and one-half kilometers out from land, or four and one-half kilometers from land.

"That's where they were to dump the whole muck. You have to actually look at what that means. Because when you say one and one-half out from land, you will automatically think that it's one and one-half kilometers out to sea. But it's not."

Instead, Micheál and his friends were horrified to discover in deconstructing the otherwise extensive, murky language of the EIS, it was clearly stated that they would be dumping the toxins within the bay itself!

"The bay is ten kilometers out, but they intend to dump the

chemicals"—pointing to a spot on the map much closer to people's homes—"right here. And we have the constant tide in.

"I was caught then. Didn't have any choice. I had to carry it through. Couldn't stop."

For the next fifteen years?

"For the next fifteen years."

FRIENDS GATHERED TO HEAR WHAT Micheál had to say about the fossil fuel company's EIS. Not bankers, developers, stockbrokers, or investors, but rural people who'd been living close to the land and the sea for generations.

Micheál and his friends decided against forming an organized resistance group at that time.

"I said that we must not have a chairman. We could not have a secretary or treasurer. If anyone was given an official job, they'd immediately become a target. We must avoid establishing positions of authority. No status. If we did any of this, I would not be involved.

"After all, I didn't know how I might react if someone offered me a million." Micheál grins. "No one else knew what they might do if they had bills to pay and got offered a million."

People respected Micheál Ó Seighin not only as schoolteacher and community activist, but as someone they could trust to make fair, rational decisions.

When One Voice for Erris, founded by Bishop Thomas Finnegan and later carried forth by his successor, Bishop John Fleming, called for a public meeting in early 2001, friends urged Micheál to attend.

"'Not at all,' I said. 'Spent my life going to meetings. Don't want to get involved any further with the Corrib gas dispute.'"

He did go, and he found many supporters of the bishop's pro-refinery position. Micheál's friend Seán Hannick, chair of the pro-gas Council for the West, was there. But as the night wore on it became obvious that this would not be a walkover. Fishermen and farmers kept arriving, moving to the opponents' side, until the bishop, also one of Michael's good friends, realized that he was going to lose.

"'Okay now, let's call off the meeting,' he announced. 'We'll form a committee, we'll make proposals and take a number of them to the developers and ask them to comment.'

"'Fine,' I said. 'But without reference to any decisions that this meeting might come to. We will continue to work individually and in cooperation with one another as we have until now.'

"The bishop got up and walked out.

"It's amazing how people will accept any kind of rubbish that the vested interests decide to pass out."

ENDA KENNY, FINE GAEL LEADER and TD (Teachta Dála, senator in the Irish Parliament) for County Mayo (who would later become Taoiseach, or prime minister of Ireland, and who is in 2019 still the Mayo TD), paid several visits to the imprisoned Rossport 5. At the same time, says Micheál, Kenny called a retired union leader named Pat, one of the five men's most ardent supporters, to say that he, Enda Kenny, would soon be sworn in as prime minister.

Micheál relates the story: "'As you know, Pat,' said Kenny. 'I'm TD for the Rossport 5's area, and I know what's going on there. You are being led up the garden path by those people.'

"Later," continues Micheál, "Kenny'd come in to see us in jail, but now he's telling Pat there should be no worry about Mayo's future, that our children have no worry at all. He says this whole thing—the opposition to Shell—is an IRA plot to destabilize the government."

Did Enda Kenny really believe that?

"Oh, not at all," answers Micheál. "Not at all. Our union friend comes in to see us, and I ask if he knows Kenny. 'Never saw the man in my life,' he answers.

"Now, Pat is always writing letters to the news media, and a few days after he publishes one letter the phone rings, and it's Enda Kenny calling again.'

"And Kenny repeats everything he'd said in the first call about the Rossport 5 having no real complaint, and that Micheál and his friends were really leading Pat astray, because this whole thing is an IRA plot.

"And then Kenny, who'd obviously been in contact with Shell in the meantime, says: 'And anyhow, I don't know why Shell is bothering with this thing at all because there's hardly any gas out there.'

"When Lorna Siggins was writing her book *Once Upon a Time in the West*, she asked Pat if what I'd told her about Enda Kenny, as I'm telling you now, was accurate. Pat said it was. She wanted to include this information in her book. But the publisher's attorneys ruled against doing so."

Michael shakes his head. "The bishop and the parish priest should have known better," he says. "The refinery's supporters never seemed to have learned anything. They hadn't a clue. They had no intention of learning anything. They didn't try to get reliable information, didn't bother educating themselves. They took the need for a refinery as the word of God, and trusted these oil people who were in positions of authority.

"Shell's people tried to come onto people's land, and we demanded that they show the legal authority for doing so. They didn't have it. We stopped them again and again from coming onto the land. Later in court they said we threatened Shell's workers, but that was never true.

"They even had video cameras and tape recorders," he points out. "There was no evidence that we threatened anyone.

"Politicians supporting Shell introduced a statutory instrument, which gave the company permission to take over people's land. This was totally outside of the law, yet the Minister of Natural Resources approved the order."

Shell went to court seeking an injunction that would prevent five named individuals—apparently those they deemed most threatening to their success—from interfering with the company's plans. All five—Willie Corduff, Philip McGrath, Vincent McGrath, Brendan Philbin, and Micheál—disobeyed the injunction.

Shell asked that the resisters be jailed.

Judge Joe Finnegan stated that he didn't know whether he was in the right or wrong, but he was nevertheless going to jail the Rossport 5 for an indefinite length of time, as well as to grant Shell the right to build the pipeline. If he were to be proven wrong on the pipeline, he claimed, he would "make them dig it up again."

Judge Finnegan wanted to keep the men in prison. However, their barrister informed the judge that he did not have the legal authority to keep them locked up indefinitely. They served ninety-four days.

WE TALK ABOUT THE RELATIONSHIP between power and fear: the power that corporations have to frighten publishers into not publishing a controversial book; the power of so-called conservatives to undermine environmental protections; the ability of organizations like the (U.S.) National Rifle Association to coerce politicians into voting against gun control legislation; the legal right of corporations to pump carcinogenic substances into the planet's water, air, and food supplies.

When the late Daniel Berrigan traveled to South Africa at the height of the apartheid regime, opponents of the government asked him what might happen if they spoke up. Father Berrigan, who spoke openly about experiencing fear when participating in nonviolent ("divine obedience") actions, answered that he didn't think that was the right question.

The real question, said Berrigan, is "What will happen if you *don't* speak up?"

IT'S LATE AFTERNOON. TERENCE CONWAY is waiting to drive me home from Micheál's. Indefatigable, Micheál tells a story about a reporter for the *Chicago Tribune* who was thinking about doing an article about the campaign against Shell Oil in Mayo. It would be a great way to inform the Irish-American community in Chicago about this struggle. But the journalist reneged, warning friends that no one must ever know that he'd agreed to write about this struggle. If certain people did hear about it, said this man, he'd never work in journalism again.

AND THE CONSEQUENCES FOR WHAT Philip and Daniel Berrigan called "speaking truth to power"?

When I was working on *Waiting for an Army to Die*, my book about the U.S. government's use of chemical warfare in Vietnam, my phone was tapped and jammed, mail arrived torn open, and Dow Chemical's scientists showed up on radio and television programs to do their best to debunk my findings shortly before I was scheduled to appear.

Vietnam veterans thanked me for caring about them, but warned that the government and chemical manufacturers of Agent Orange were not going to allow this book to be published.

They were frightened for me. *The truth must never be told about the massive use of toxic herbicides in Southeast Asia, a flagrant violation of international treaties forbidding the use of chemical weapons in warfare. If I insisted on writing and publishing this book, they feared, my children would grow up without a father.*

MICHEÁL'S OWN PHONE HAD been tapped for years.

"There was one very bad time," he recalls, when he had "lots of trouble with the phone until I decided to tell the repair technician who came by that the phone was tapped.

"He came back a couple of days afterward, and said, 'Okay, that thing about your phone, it won't happen again.' He didn't even pretend that it wasn't tapped."

I ask Micheál about the presence in Mayo of mercenaries such as those employed by multinationals and governments around the world to intimidate and even kill local people who object to the theft of their communities' natural resources.

The enforcers Shell brought to Ireland were ex-army men, hired to harass and beat the company's opponents. They videotaped children changing into swimwear and playing on the strand. A man from Dublin living in Mayo for a time got upset because mercenaries were always watching his children, so he complained to the police superintendent.

Micheál recalls the result: "'Oh, now, don't worry about it,' says the superintendent to the man. 'Next time I'm with Shell I'll talk to them about it.'

"I remember one of the men Shell brought in," Micheál continues, "a rather friendly black man from South Africa, who said that he was here in Ireland for a rest. He'd been in Namibia and fought in Angola with the South African strike force. When Mandela came to power he got out of the army, but then he spent time as a mercenary with the Serbs in the attack on Bosnia. After Bosnia he worked in Iraq for the mercenary company Blackwater. He was not perfectly normal, but you can imagine the kind of head he had after all the kinds of things he'd been through."

Complaining about these mercenaries did no good.

And it got worse.

"ONE OF OUR OWN IRISH POLICEMEN," says Micheál, "even told John Monaghan that he was going to rape John's wife, Brid, who happens to be my own daughter. John recorded this threat. He complained to the ombudsman. An ombudsman came down to talk with John."

Micheál places a map of the region on the table to point out where John and Brid reside.

A police officer threatened to rape your daughter?

"Yes, a sergeant in the guards. After that, this man started driving in at different times of the day, himself and a mate going past John's house. He'd stop the car and then back up and wait. Or he parks the car up at the house and leaves it there. Goes down the road for a while, comes back again and moves the car."

The two men assigned by the ombudsman's office to this case were standing beside a window in John's house when the sergeant and his mate drove by, stopped outside, drove on a bit, stopped, and then backed up.

Micheál relates the conversation that ensued: "And John says to the ombudsman's men, 'I've been to the ombudsman's office in

Belfast, which is where complaints can be filed, not just against the police. But so far as I know this office investigates, but does not discipline police officers.

"Next," says Micheál with a snort, "the guards decide to 'punish' this sergeant for threatening to rape my daughter—by *promoting* him from sergeant to inspector."

Micheál would not be silent. "Here," he says, "is what I told the superintendent and the guards: 'If this man hurts anyone in my family, I will kill him. You have shown that there is no point in complaining. You will not investigate John's charges. Remember. Just remember. I will kill him or I will see him killed.'

"And they knew I would do that. No, the man did not try to get back at me. I broke the law openly by threatening a guard. Now that makes me a complete, absolute criminal. I have said that again and again and again, but they haven't touched me. And they won't.

"It's a disgrace that you have to be prepared to do something like that in what's supposed to be a democracy, in a country that had to fight very hard for its independence," Micheál says. "We have lost to a large extent the sense of obligation to identity."

And the ninety-four days he spent in prison?

"It was fine," Micheál recalls. "I knew why I was there. I broke the law, but many of the prisoners told us that we'd done nothing wrong. We did not belong in prison at all.

"I told them, 'We did the same thing as ye did. Ye broke the law, we broke the law. A judge decided that we had broken the law, so it was perfectly true we belonged in jail.'"

WHILE HE DIDN'T ACTUALLY *ENJOY* the experience of being locked up, it was not as dreadful as one might imagine. The men were given six hours daily to exercise in the prison yard and spent the rest of the time in their cells or at meals. "We had an electric kettle in the cell to make tea," says Micheál. "We had television. Prison food, prepared by the inmates themselves, was unbelievable, just fantastic."

He was serious. Plus, each day, people from around Ireland—and the world—came to visit the Rossport 5. The men were being kept in a holding prison, for inmates waiting to be charged, rather than in a long-term facility. Things were quite different, they knew, than they might have been elsewhere.

"Most of the warders supported us, and they hated the Minister for Justice, Michael McDowell, a grandson of a founder of the [historically pro-British rule] Irish Volunteers. All of the prisoners hated the Minister for Justice.

"It was miraculous," recalls Micheál, "because all of the nuns and some of the [members of Catholic] contemplative orders in Donnybrook and Dublin sent a card signed by all the nuns. Here we were in the middle of the Celtic Tiger [economic boom], when everyone was concerned with money, more money, and you had these [us] men from Mayo who just said *No*. The musicians supported us because I would be fairly well known among them, and Vincent McGrath would be awfully well known. The first day we were in jail, a piper turned up to lament us outside the jail. He came every evening . . . until we told him to go away!

"There are really three kinds of people: One, those who believe they are going to benefit financially from a project like Shell's refinery. Two, others who, like us, looked into this project, studied it, came to understand what's involved, and decided to oppose it. And third, there are *most* people, who worry about something like this but aren't motivated enough to do anything."

"[JUDGE] FINNEGAN DIDN'T WANT to let us out of jail," Micheál says. "He would whine, 'They have not purged their contempt of me.'

"Of course," Micheál laughs, "the authorities didn't expect us to stick to our resolve. Here were five 'thick' people up from the country. They thought we'd never be able to stick it [out] among drug dealers and teenage scamps. After a while, the authorities started to complain to TD Jerry Cowley. 'If only for Ó Seighin, those men would be out,' they said. 'If only for that Shine.'

"Finnegan said that a judge would be available [to rule for our release] twenty-four hours a day—all we had to do was to 'purge our contempt,' to say, 'Sorry, we won't do it again.'

"Even our barrister, an excellent man who had at one time been Attorney General, said, 'Now, if one of ye went out to purge your contempt we could then take a High Court case.' We said to him, 'Look, will you please have sense, man? If one of us purges our contempt, we will have broken the momentum. It won't be done.'

"You see," Micheál explains, "Shell and company look on us as native people, and there's the same attitude to us that they have toward Native Americans on their reservations. The same attitude they'd have toward Nigerians. You know, like we're all primitive beings who don't know what we're doing.

"And they still feel that way about us, in spite of the fact that we have hammered them in public forums. We have defeated the best of their technicians. We have defied them, beaten them down. We have beaten them in public debate.

"We forced the County Board to look at reality in a new way. That is, that the mathematical reality of an engineer is not the only reality. It's just one aspect of the analysis of reality. *We know* the reality. *We know* the bogs. *We know* the water. *We know* how the community relates to one another. *We understand* that some people are greedy and others are not. Shell found out fast that *we* are the pragmatists. We *will not tolerate* the pushing of theory on us, the pushing of nonsense on us.

"I will not take bullying. I will *not* take it. Bullying is number one on the lists of the eight deadly sins.

"Even so," he says, "I believe they still have this attitude that people here are an ignorant kind of primitive thing who are totally against development, that we can't understand the 'wonderful' modern world.

"There was an old musician who'd had strokes, and he hadn't played any music for a long time. During the first march in Dublin, he needed to be helped up to the stage to play the flute. This man came in to see me in jail, and he said: 'Now Micheál, this is the most liberating thing since 1916.'

"I was on a radio program one morning and they were talking about all the opportunities Shell was bringing to Mayo, you know, *the Jobs, the Jobs* are everything. And I said, 'Wait a minute. Jobs are not the most important thing. The health of the people is the most important thing. Anything after that is secondary.' That was quoted again and again and again."

SPONTANEOUS CELEBRATIONS BROKE OUT throughout Ireland the day the Rossport 5 were released. Thousands of people marched through Dublin. Revelers also included the members of the workers union at Waterford Glass, the internationally famous glassware manufacturer, who had picketed on behalf of the Rossport 5 every single night after work.

Shell realized it had made a mistake by failing to read Irish history and refusing to believe that ordinary people can and will withstand sustained punishment.

AFTER FIFTEEN YEARS OF STRIFE, divisions, intrigues, police brutality, mercenaries, jail sentences and prison terms, is the community beginning to heal?

"No," Micheal says. "There's a much better chance of healing when people are still mad at one another than afterward. There is no healing. There will be no healing. Remember, neighbors betrayed their neighbors."

Shell to Sea became the key grassroots group opposing the project

Communicator

> Maybe our constitution needs to be rewritten to more accurately reflect the reality of modern Ireland. Maybe it should say: "The State is bound to protect the profits and self-proclaimed economic entitlements of every industry big enough to hold sway over government policy."
> —Valerie Finan, Save Our Cancer Services-NW campaign, in a speech to the Martin McGuinness public meeting held in County Sligo, October 2011

IN THE EARLY YEARS OF THE SHELL CONTROVERSY, Mark Garavan attends small, rather quiet meetings that in time will grow into raucous gatherings and large protests. Homeward bound on narrow dark roads, he thinks about Shell's violence in Nigeria. *How far might this company go to crush opposition to its plans in County Mayo?* Mark knows he is being watched and his phone is tapped. Shell Oil's mercenaries seem to be everywhere.

Mark tells his wife that if he should be seriously injured or killed out on the road, it will not be an accident. A skilled, careful, sober driver, he is most unlikely to die in a car crash.

IT'S A FINE FALL DAY AT THE GALWAY-MAYO Institute of Technology in Castlebar. Students are chatting on their phones, greeting friends, telling stories, joking, and laughing.

Young men and women hoping to graduate from college, find jobs, start families, and live long, happy lives. They believe in the future. After all, Ireland has survived centuries of colonialism, rebellions, poverty, civil war, and the Great Hunger. Thousands of years from now, this small green island will surely still be thriving.

Have they read Pope Francis's 2015 *Encyclical*, in which he warns that time is running out to save the planet from irreversible harm to its ecosystems? Are they aware that Francis quoted fourth-century cleric Basil of Caesarea in calling unfettered pursuit of money "the dung of the devil"?[1]

I am here to meet faculty member Mark Garavan, editor of *The Rossport 5*, an excellent oral history of the price five County Mayo men and their families paid for resisting Shell Oil's attempts to usurp the rights of citizens that are enshrined in the Irish Constitution.

Mark's story begins in a classroom at the Galway-Mayo Institute of Technology, where he's teaching a course in Tourist Guiding. It's 1999–2000. Professor Garavan and others who love Mayo's beautiful coastline are working on ways to preserve "the Wild Atlantic Way," by developing that region for ecotourism. Students are excited about this idea, but they are worried about a project that three companies—Enterprise, Marathon, Statoil, and later Shell Oil—are planning to construct in Mayo.

No one knows for certain what these fossil fuel companies might do, but surely they would not dare to mar the unspoiled beauty of western Ireland.

After much thought, study, and debate, Mark's students decide to express their objections by writing to the Mayo County Council. They want to create a dialogue with supporters and opponents of the refinery. Isn't that the way things work in a democracy? Citizens express their views. Others express counterviews. Experts conduct studies, weigh evidence, and make decisions based upon careful analysis of reliable information. No one person makes decisions for the majority, and it's always possible to appeal flawed or dangerous decisions.

"Many people felt that this thing must absolutely go ahead in Mayo," Mark explains. "'If it does not come to Mayo,' they warned, 'then County Sligo or Galway will grab it.' A terrible loss for Mayo, whose politicians were jumping up and down, assuring everyone not to worry, Mayo was definitely going to get this windfall.

"Mayo, say cheerleaders like the clergy, will resemble Saudi Arabia. Sure, there will be so much money this area will turn into El Dorado.

"Now, you might think I'm exaggerating," Mark continues, "but all this is being said, all the time: 'Everyone is going to be rolling in money.' When asked how people are going to acquire this great wealth, it is much less clear. The whole thing was ridiculous."

> In 1956, I was with my father when crude oil was found at eleven o'clock in the morning. We thought that when oil was found—we people out here are very poor—we thought we would be millionaires. We are still depressed. The town is very tattered. Shell promised to build schools and to make a sea wall because the town is flooded every year. Nothing was done.
> —EDWIN OFONIH, TAX COLLECTOR IN OTOIBIRI, NIGERIA[2]

GERARD AND MONICA MULLER, SOME OF the first local people to show real concern about the refinery, gather with friends and neighbors to talk about the project. They urge everyone to read carefully the oil company's Environmental Impact Statement (EIS). People write letters to government agencies and to the three oil companies initially behind the Corrib gas project.

Students living in the coastal region are also concerned. This refinery is totally at odds with ecotourism. It will be terribly destructive, they are convinced. Mark finds it hard to believe that anyone would even consider building something so totally incompatible with Mayo's undeveloped, pristine coastline.

"What a stupid, contradictory idea," he recalls, "that we would be developing strategies to promote ecotourism along the coast, and at the same time agree to support building an experimental gas refinery right in the middle of ancient beauty.

"It was all so bizarre, like complete madness. This was our opposition's starting point."

Mark's students file objections with the Mayo County Council. *Under what circumstances will this project go forward? Who will benefit? Can this development possibly be legal?*

"And then," says Mark, "I started looking at the planning files, discovering all sorts of strange things—lots of evidence of illegal activity in terms of how the planning process is being conducted.

"From the beginning, the process of granting consent for the refinery is split between different agencies, the main body being Ireland's Department of Marine, the minister of which, Frank Fahey, is very much in favor of this project. He is *very* partisan, a strong advocate of building the refinery. Yet at the same time, he is supposed to be the *independent regulator*. That is, the one authority deciding whether consent will be given to allow the project to be developed."

Fahey issues something called a "foreshore license" that will permit the project's developers to do almost anything they want, up to the foreshore. But for this case the Ministry defines the foreshore as being 8 kilometers inland, which Mark knows is completely illegal. The foreshore is the part of the shore between the high-water mark and the low-water. That's it.

"We went to a public seminar," recalls Mark, "at which [a representative of] the Department of Marine said, 'Oh no, we have special exemptions to do this.'

"The whole thing was cowboy decision-making. They had no respect for standard process because they figured that everyone was in favor of the refinery: Everybody wanted it to happen, nobody would possibly object."

He points out that historically, Ireland has not wanted to appear backward, and has taken most available opportunities to participate in modernization. "When Ireland found gas, for the first time since the 1970s, people thought this would be a whole series of finds, and the government wanted to make sure exploration continued. Any opposition to all this would be giving the wrong signal, as though Ireland wasn't open for business."

IN AUGUST 2001, MAYO COUNTY COUNCIL grants permission to build the refinery. Local people appeal, and the Planning Board holds a hearing in November 2002. Lasting twenty-two days, this will be the longest such hearing in the Board's history. It turns out to be an extraordinary event because Planning Board inspector Kevin Moore makes a statement that will be widely quoted by Shell's opponents in the years to come:

> It is my submission to the board that, from a strategic-planning perspective, this is the wrong site; from the perspective of government policy, which seeks to foster balanced regional development, this is the wrong site; from the perspective of minimizing environmental impact, this is the wrong site; and consequently, from the prospective of sustainable development, this is the wrong site.[3]

Moore is a good public servant. He pays attention. He listens. He allows local people to speak in their own voices. He allows all local views to be heard, and it is an incredible hearing. For the first time, people begin to listen to two sides of the argument, and speakers give cogent, coherent, logical reasons why the refinery should not be built.

The chief inspector finds in favor of the local community, and he turns the project down on a whole range of grounds. This is a very big deal. It causes a political crisis behind the scenes. The Irish government's reputation is now on the line.

"To understand where the government was coming from," Mark

says, "you have to realize that our economy since the 1950s has been built upon attracting foreign investors. We have no real indigenous base, no industrial base, so the Irish government developed a very aggressive policy to attract foreign investment, primarily U.S. companies to use Ireland as their own base of operations.

"It really got going when we joined the European Union because companies could then export to Europe from Ireland. We had very benign low taxes, to attract companies coming in. So, developing oil and gas was seen in the context of these broader policies.

"The government wants the oil and gas companies to come to Ireland offshore. It wants to explore the marine environment. Whatever the corporations want, that's what the government *must* do. We're servants of the corporations.

"Now, we know that politicians at the very top level are meeting with senior executives from Shell and giving commitments that the Corrib project will go ahead, no matter what. That is a matter of record. These guarantees come straight from the government to Shell. And even if officials have to change laws, that's exactly what they are prepared to do."

To Shell and its friends in high places, the Corrib gas project is a done deal. The refinery will be open, fully running, by 2005. But then the company makes a serious strategic mistake—sending five Mayo men to prison for refusing to obey an injunction allowing Shell onto their private property.

Local people have been objecting to Shell's pipeline and refinery for five years. Now, for the first time, this resistance becomes a national issue.

"This whole thing is suddenly a very public event in Ireland, and the first major protest will be held right here in Castlebar. We expect about two hundred people to turn up, but instead thousands come out. Things have changed. The resistance has gone national and we realize it will need a new direction."

Large crowds gather in Sligo, Galway, Dublin, and other places. Public meetings are held throughout the country, and Mark agrees to act as spokesperson for the Rossport 5. Members of the board for the school where he teaches try to convince him to end his involvement with Shell's opponents. Now, massive crowds turn out to hear Garavan and others talk about why the refinery must not be built.

But why would sending five men to prison generate so much resentment? When ordinary citizens are jailed in the United States for opposing dangerous gas pipelines or other fossil fuel projects, the media pay little attention. Large protests, like the one against the Dakota Access Pipeline, draw national attention for a time, but the press soon moves on to other stories.

Mark doesn't think the huge protests throughout Ireland were really about five ordinary men going to prison. People wouldn't necessarily care very much about that. The strong reaction results from a combination of things.

"First, it is about who the Rossport 5 really are," says Mark. "The media can't categorize them as crazy people, eco-warriors, or radicals—standard terms the media uses to disown or diminish individuals and to denigrate the cause they may be advocating. These are five ordinary guys, local people, and middle-aged men, not just 'protesters.'

"Second," he continues, "sending people to jail in Ireland for defending their own land is a big cultural trigger. It reminds the Irish of colonialism; you know, nasty English landlords, and centuries of Irish people standing up for land rights. Many people remember Michael Davitt's Land League, a very big movement in Mayo back in the nineteenth century when tenants fought landlords for control of land.

"Not that long ago, big landlords owned all of Ireland's land. Everybody else was a mere tenant. So, in this country large powerful people putting small farmers in prison has a huge historical trigger.

"A third factor is the dramatic scenes on the streets of Dublin outside of the courtroom. The jailed men's wives are there, speaking clearly, without jargon. They do not have a political agenda. They are emotional, incredibly articulate, and quite moving. Many people identify with these wives and mothers. Mary Corduff, in particular, is a powerful presence."

Speakers for the campaign present a clear message, explaining exactly why they support the Rossport 5. *Here are five good men, standing up for rights enshrined in the Irish Constitution. And a multinational corporation is violating those rights by putting them in prison.*

Shell doesn't appear to know Irish history. Nor does it know, or care to know, the Irish people, which is why it undermines its own public relations campaign by appointing the haughty Andy Pyle as managing director for Ireland. This man speaks with a heavy English accent. He cannot, or he will not, pronounce basic Irish names and the names of places, like "County Mayo." Willfully, it seems, he continually mispronounces the imprisoned men's names.

From an Irish point of view, this guy is an evil ogre—the wrong image altogether for Shell, but in a sense gratuitous because it reinforces the five men's motive, and helps people relate to the prisoners and their families, who are obviously being oppressed by some huge English-controlled company.

As Ike Okonta and Oronto Douglas write in their 2003 book *Where Vultures Feast:*

> Shell employs a sophisticated array of damage control experts, scenario planners, lobbyists, and spin doctors to present the image of a caring, thoughtful, and socially responsible company to the outside world. Long before the issue of the environment became a topic of national discourse in Europe and the United States, and multinational oil firms were forced to adopt the veneer of environmentally friendly companies, Shell had elevated the concept of selling itself to the powerful conservationist lobby into an art form, devoting a considerable chunk of its budget to this effort over the years.
>
> So successful was this portrayal, and so skilled were Shell's image makers in contriving a semblance of "constructive

dialogue" and openness toward critics, that it took a major public relations disaster like the company's intention to drill for oil in a protected forest in rural England in 1987 for this carefully cultivated image to shatter and for conservationists and ecologists to see Shell for what it really is: a modern-day Gulliver on the rampage, waging an ecological war wherever it sets down its oil rig.[4]

"SO," SAYS MARK, "PEOPLE ARE ASKING, 'Do you want to live in a country where a big company can come along and put something on your land that might injure or kill your family? And you will have no say about all this?'

"Then there was the question about whether Ireland might benefit hugely from Shell's plans, and you might say, 'Well, maybe it's okay because overall we're benefiting.' But then you realize that there will be no benefit from all this. None."

IRELAND HAS BEEN RANKED AMONG the world's 15 worst corporate tax havens, according to the 2016 Oxfam report "Tax Battles." Oxfam reveals that the "tax havens are leading a global race to the bottom on corporate tax that is starving countries out of billions of dollars needed to tackle poverty and inequality."[5]

"THE FIRST TIME I ACTUALLY MEET SOMEONE from Shell is on *The Late Late Show,*" Mark remembers. "I'm scheduled to debate Shell's chief executive for Ireland, Andy Pyle, and he appears to be completely shaken. Before we go on air he won't look at me, refuses to speak to me at all. He's in a dreadful state. I don't think he'll be able for it. I realize then that Shell is broken. They are desperate and just can't keep this whole thing going."

At the height of the turmoil, Mark is invited to speak to a group called "PR in Ireland." The event starts out as a debate with a particularly obnoxious journalist who urges the audience "not to be fooled by Garavan's nice soft voice. This man," warns the journalist, "is malevolent: a thug and a manipulator."

Many people in the audience happen to work for state companies and private companies. To Mark's surprise, one of the corporate employees confides that since 2005 his company has studied the resistance movement's Shell to Sea campaign. "Brilliant," he says, admiringly. "You are absolutely brilliant. Your campaign managed to get five men into prison. Then you told their story to the world. Brilliant public relations. Unbelievably good."

The man goes on to cite the campaign's great speeches and rallies, how everything appears to be designed to destroy Shell, how just when the company begins to maneuver against resisters, these clever people internationalize their story, putting out information about the terrible things Shell has done in Nigeria. Then Willie Corduff wins the Goldman Environmental Prize. *That* is a really big deal.

"And as this guy is saying all this," laughs Mark, "I am thinking, *He doesn't have a clue.* None of what he is talking about was ever planned. He should come to our meetings. They are chaotic! Zero structure, no coherence. Most of the time we are roaring at each other instead of at Shell.

"I mean, you couldn't find five more different men than the Rossport 5. Put these guys in the same room and they would never necessarily agree about anything. How all this worked together is an untold miracle. We didn't pick the Rossport 5. Shell did that for us. They could have arrested loads of people, but they chose not to. Moreover, the international publicity had nothing to do with us. [Executed Nigerian environmental activist] Ken Saro-Wiwa's brother, Owen, did come to Mayo several times, and of course we always welcomed any outside support.

"I couldn't say all this at the time, with these people looking in at us and thinking we are a coherent, organized, always-in-agreement group—which was really the opposite of how things worked. We never even had posters or banners or advertisements. [There was] no money for taking ads in the local papers. All voluntary work, all done on a shoestring, with hard work . . . and lots of luck."

People appear at Shell to Sea meetings with all kinds of advice, proposals, and ideas about how the campaign should be conducted. One man who hasn't really been involved announces that the resisters need to develop a class analysis. Without that, says he, the campaign will not succeed.

"And the people in the campaign always listen with courtesy and respect to anyone who comes to those meetings," marvels Mark. "They are incredibly patient, incredibly hospitable, happy to have the support."

AFTER THE FIVE MEN ARE RELEASED from prison on September 30, 2005, the resistance enters what Mark calls "the delicate stage."

Like a sports team that assumes it is world-class professional but keeps losing games, Shell decides to reassess its game plan. Suspending the project for one year, the company buys time to devise new strategies to assure the refinery will be built.

Shell needs to get the local media on its side. It hires recently retired police officers as consultants, and acquires the services of well-known Mayo journalists to write favorable stories about Shell. Securing the support of local businesses, company leaders know, is imperative to this new campaign.

Shell's opponents begin hearing the results of this propaganda campaign wherever they go. Confides one Shell-friendly businessman to Mark, "You know, we have a text alert. Every time we hear you [Shell's opponents] talking on the radio, we're supposed to send in a text complaining."

Shell encourages listeners to flood radio stations with hostile reactions when they hear someone who does not support the refinery. A lot of media is brought over to Shell's side, particularly journalists working for the *Irish Independent*.

"You don't even have to give some journalists a gift," says Mark Garavan, "because they are just so politically demented. You know, like one story in a national newspaper claims that an IRA unit is operating in north Mayo. That these subversives are living in the mountains and come down to villages at night, going door to door, intimidating people, ordering everyone to oppose the refinery. It's pure fiction, right out of a novel. And there are lots of stories like this in so-called 'reputable' publications."

Shell wants Westport taxpayers to stop caring about the Rossport 5. They want the farmer in Kerry and the taxi driver in Dublin to stop caring. Thus the company must find new ways to marginalize resisters. It needs to portray them as radical protesters, 'eco-warriors'—odd, outside the mainstream, totally crazy. This will weaken the campaign itself.

At the same time, the company will transform its own image from a greed-driven invader into a reputable, benevolent corporation endeavoring to bring good things to Ireland:

> We regularly discuss project impacts, including security, with individual residents, local groups and community representatives We are committed to ensuring that our construction works and associated activities have the least impact possible on our neighbours and other members of the community.[7]

Shell replaces Andy Pyle, the British managing director for Ireland, with Canadian Michael Crothers; brings in local people with Irish accents to promote the project; and pays off a lot of people, handing out funds to companies in north Mayo, giving sheep pens to farmers. One businessman reveals that Shell is offering him thousands of euros to develop his business. Shell is openly doing all kinds of things it claims it never does—buying influence and favor.

Now Mark laughs: "Shell calls all this paying off they're doing 'community liaising,' or 'community sponsorship,' 'community building.' You know: 'Look how good we are.'"

Shell's cleverly designed public relations campaign obscured the truth about why local people oppose the refinery. Mark and his allies were gobsmacked by the range of the fossil fuel company's resources. "They did begin to control the media," says the man charged with leading the communications campaign for the opposition, "making it nearly impossible for us to be heard.

"The access we'd had when the Rossport 5 were in prison was total, but now it was very limited. You'd have a lot of law-and-order kind of narrations, which helped Shell a lot."

Mark explains: "A huge police force is brought in, not well trained, if trained at all, for this kind of community conflict. Very aggressive police, increasingly antagonistic toward nonviolent protesters. They know nothing about how to manage civil protests. Being paid in some liaison capacity, these police are acting hand-in-glove with Shell's mercenaries.

"I used to give this analogy whenever I spoke about the Corrib gas project: Let's say I offer you a winning lottery ticket. Guaranteed two million dollars, and you say you're not sure if you want it. In County Mayo, politicians were behind the refinery, the Church was behind the refinery, and 99 percent of the people were behind the refinery early on. And here's this one percent who keep talking about not wanting this jackpot lottery ticket.

"When you are at odds with national institutions—the media, the Church, mainstream political parties—when you are refusing 'great opportunities' for your community, great sums of money, then people will say you're insane and will not listen to anything you have to say."

"AND NOW, YOU SEE, THE REFINERY *is* here, the gas, and *no one is benefiting one iota from it,* which is exactly what we've been saying all along. The arguments that our campaign was making have been proven empirically, demonstrably, correct. There's no disagreement about that.

"After so much turmoil and pain, Shell is gone, they sold their interest in the refinery. Who knows if they made any money from all this trouble, but for them it has been a complete public relations disaster. They're not stupid people. They know what happened.

"One hard lesson from the campaign was learning that many politicians cannot be trusted. Politicians acted like small children wanting Halloween treats: close to home one minute, chasing off after new opportunities the next. [They'd be] offering to help Shell to Sea, supporting the Rossport 5, making many promises, talking a lot whenever a crowd gathered, assuming they might gain support. Then, once elected to office, turning their backs on Shell's opponents, abandoning them.

"They completely sold us out," Mark states. "Not somewhat, not halfway, but completely. Their support for us evaporated. It was appalling. They were treacherous."

He adds: "An important footnote here. The Rossport 5 won in court. Shell's injunction had, in fact, been illegal, and now the men have been proven right in their defiance of it. They should never have been sent to jail."

Officially, Mark says, Shell claims it left the Corrib to "consolidate its portfolio." But, he points out, it was always "a dumb idea, building a refinery on land and then connecting it to a very long pipeline. Shell played hardcore power games in County Mayo, but ultimately in some very important ways the company was humiliated and defeated. Grassroots resisters fought Royal Dutch Shell to a standstill more than once."

"County Mayo people tried to tell Shell: 'This is where we live, where our fathers and their fathers before them lived. It's our place, and we don't have room for this big industrial project. We want to keep our relationship with this beautiful region. We do not want *place* to become a threatening thing.'"

As Mark notes, many people were traumatized during and after the Shell to Sea campaign. "In hindsight," he muses, "we should have been kinder to one another. The meetings were very tough, and we needed to take more time to look out for one another.

"People were under tremendous psychological pressure. We had no time to stop and take stock of what we were doing, no time for a weekend retreat. Just intense day-to-day pressure. I'm willing to talk about how to avoid some of the difficulties Shell to Sea faced, and to share the lessons we learned. We want to share our knowledge with other campaigns, and they will pass on what they've learned, or are learning right now, to us.

"We defeated Shell on all substantial issues, even though they completed their evil project. Some people say, 'You just have to go ahead and accept it.' You know, as though you've been diagnosed with cancer, so just sit back and wait to die. We are never going to do that."

Finally, says Mark, "I have not seen the refinery. Can't drive by there, couldn't look at it. That would just be too painful."

IN THE DAYTIME, THE REFINERY APPEARS innocuous. There's no loud industrial noise, no observable pollution. Inside, ordinary men and women are working to secure food and shelter for their families, though Shell did not provide most of the jobs it promised.

But at night, on the road from Ballina to Kilcommon Lodge Holiday Hostel, you will find bright lights rising out of dense darkness like an alien spaceship. It's not a hotel or resort, not a little village. On one side of the road there's a row of crosses, on the other a fence, with the entrance to a gas refinery. Something that has not been here always. In time, aging pipes will break, buildings collapse, gates tumble from their hinges. Stories will be handed down, songs and poems written about a people's uprising in County Mayo.

Fritz Shultz facing police in Ballinaboy, County Mayo

Innkeepers

> Shell is committed to transparency as it builds trust. Trust is essential for a company that operates in our line of business, reflecting our core values of honesty, integrity and respect for people.
> —Shell Oil, Public Relations Material

BETTY SHULTZ WELCOMES ME INTO the common room at her Kilcommon Lodge Holiday Hostel and begins by giving me some background. Times were tough in County Mayo when she and her husband, Fritz Shultz, arrived from Germany in the early 1980s. People were leaving to find work in England and America. Not wanting to take a job from anyone, the new immigrants hoped to create something new and vital, something to share in the place that had given them a very warm welcome. They renovated and extended a 200-year-old building in Pullathomas and, in 1984, opened their hostel, surrounded by a stream, gardens, and mature trees.

Born in 1957, Betty grew up on an isolated farm in postwar northern Germany. Her parents went shopping twice a year, and the family was otherwise self-sufficient, making almost everything themselves.

"My mother was the local district nurse," says Betty. "I knew little about my country, just our farm. We didn't even have television until I was twelve, and my dad allowed us to watch one hour a week. So we had little knowledge about the rest of the world.

"I left home and started my nurse's training; it was an eye opener seeing many patients with amputated limbs and other serious injuries, a very impressive reminder of what had happened not that long ago."

During the war, Betty's mother had lived in Dresden, one of the German cities destroyed by Allied bombing campaigns. She told stories about surviving fire-bomb raids and losing her home, as well as a good bit of her identity. She was struggling to regain certain parts of herself.

Betty's maternal grandfather, a German general, returned home in 1955 on the last train carrying prisoners of war from Russia. He had been in Russian prisons for many years, his family never knowing if he was still alive.

"My grandfather came back to a very changed country," Betty says. "My mother was about to get married, and soon her children were born. I have good memories of my grandfather as a grandfather. He was playful, a great storyteller who always took us to the beach. I saw the many deep scars and wounds on his body, one on his back so deep I could have put my whole hand into it. And the adults would be in the sitting room, off-limits to children, talking about serious things, like living with the new reality of Germany."

A soldier in both world wars, Betty's grandfather died when she was eleven, which was unfortunate because Betty wanted to ask him many questions.

Some of Betty's schoolteachers clung to Nazi ideology, while others assigned the writings of concentration camp survivors. Suffering from a severe case of post-traumatic stress disorder, one teacher would suddenly shout that the school was being bombed. He'd order students to crawl out of the room and into a shelter. Not wishing to upset this troubled man, they did as they were told.

Coming out of this flashback a short while later, he was always terribly embarrassed.

Fritz, born in 1933, carries memories of living in Hamburg through saturation bombing campaigns, his own house being destroyed by British bombers. The family fled to the countryside, arriving on the doorstep of a farmer, who turned them away.

BETTY AND FRITZ HAD BEEN LIVING in Mayo for more than twenty years, raising four children, nearly totally self-sufficient, when people began talking about the Corrib gas discovery.

"We'd heard all kinds of things like this before—someone was going to use the tidal energy from the bay, someone else was going to dig for gold in nearby cliffs," recalls Betty. "You hear all these things.

"We took the children to the beach at Glengad almost every day—so beautiful, peaceful. The idea of a pipeline coming through our beach sounded quite silly. And a refinery in the little village of Ballinaboy? That was hard to imagine, it didn't make sense. In the beginning, we just didn't believe it.

"Then Shell applied for planning permission to build and it seemed actually true; they wanted to construct a refinery. We could object to the planning authorities about Shell's project, and a few people in Ballinaboy and other villages in the vicinity did do that. They were the only people we knew who, like us, were not happy with this project.

"Our children spent a lot of time with friends in Ballinaboy, about seven kilometers from our house. And those kids often stayed with us, so we heard a lot from them about Shell's project."

Betty and Fritz submitted an objection to Shell's plans. Fritz was suffering from cancer at the time, which limited how much the family could get involved with the growing resistance to Shell. They did not yet know Micheál Ó Seighin and others in Rossport; however, Gerard Muller stopped by one day to talk about the dangers of building a refinery and pipeline in Mayo.

"Shell's first application was turned down, resulting in big celebrations in the pub in Glenamoy," recalls Betty. "One neighbor came into the pub dressed like a Shell official, and he pulled a piece of paper out of his pocket that read 'PLAN B' in big letters. We laughed because we were certain there was no Plan B.

"We thought we'd won, Shell would pack up and go home. Not everyone, of course—a lot of people liked the idea of having some kind of development here. After attending a couple of public meetings, Fritz said he didn't think there would be that much resistance. In fact, many people seemed to like Shell's plans. All of the local councilors were happy."

At one meeting, Sister Majella McCarron, a Catholic nun who had lived in Nigeria for thirty years, talked about Shell's crimes against the Ogoni people. The military junta had murdered her close friend, activist Ken Saro-Wiwa. If Shell were allowed to go ahead with its plans, she warned, terrible things might happen in County Mayo, Ireland, as well.

In the beginning, opponents of the refinery felt like a tiny minority, possibly out of touch with what their neighbors wanted.

"I remember one of our children coming home from school," says Betty, "and he said, 'Mom, will you please stop being so German?' You know, Germans inspect everything; they study everything; and they ask, 'Is this really good? Do we want this? Let me know all of the details, and only then will I have an opinion about something.'

"So, yes, jobs sound good, but how many jobs and for how long? It seemed like nobody else wanted to ask these questions. Our kids felt we were the odd ones out by warning that Shell was going to dig up everything, pollute the area. Some kids in Ballinaboy said [to our kids], 'Ah come on now, we're going to get Internet access.'"

Betty and Fritz drove to one public meeting, where they hoped to speak with Shell's project manager about their concerns, but when they arrived the local priest and some of Shell's managers were drinking together at the bar.

"We decided to not try talking with them," says Betty. "After all, if the local priest is raising pints with Shell officials, whatever we might say is going to be in vain. We turned around and went home."

Fritz and Betty's Kilcommon Lodge Holiday Hostel looks out to the estuary and, just down the road, a cemetery that rests upon a hill that collapsed in 2003, cascading graves into the water. (This is the same hill that Micheál Ó Seighin had warned his students would surely come down if Shell tried to push its gas pipeline through unstable soil; that it collapsed even without the pipeline proved Micheál right.)

"We laughed about that crazy idea," Betty says. "It was obviously impossible [to put a pipeline there]."

AFTER THE FIRST HEARING INTO THE REFINERY, Shell's project manager and his sidekick came to visit Betty and Fritz at their hostel.

"We sat them down by the fire," Betty says, "served tea, and told them, 'We are here because you are not here. We live here because there is no pipeline, and there's no refinery.'

"And they said, 'But this project might lower your energy prices. There'll be more gas, and you will need less money to heat your house. This project will create local employment. Shell's workers will fill your hostel for years and years. You'll be booked out forever. You can even expand.'

"If you employ local people," Betty told them, "they won't need accommodations because they have homes right here. You don't understand. We are not in Mayo for refinery workers. We're here for people who want to spend time in this beautiful, unspoiled, clean environment. People come here and then return again and again. What would we do without them? Even if we accommodated Shell's workers for three or four years, we would lose all the good business we've already built. And what for? Something we do not want in the first place.

"I said to them, 'We're talking different languages. My son is training to be an outdoor instructor. How does that fit in with your plan?' 'Oh,' they said, 'but we would have work for everyone in the family. Your son could even be one of our drivers.' "

"We made it clear this was the last time they would come onto our property," Betty says. "We did not want them here ever again."

Was there a turning point, a time when Betty and Fritz knew they must actively resist Shell's plans?

Yes, they say: after the jailing of the Rossport 5—men Betty and Fritz respected and trusted, the last people in the world who should have been in prison.

A neighbor called when the men were locked up to say folks were gathering at Shell's proposed construction site. When Betty and Fritz arrived, already two hundred people had gathered, with more arriving by the minute.

"We looked around and we knew all the faces," says Betty. "I was surprised because I did not know so many people had strong opinions about this thing. One of our son's teachers from Belmullet was there. It was a hot day, and soon P. J. Moran brought a sheep trailer with lemonade and sandwiches, and the local shop provided more sandwiches. Mary Horan started dishing out tea.

"I was nervous that everyone would go home—children would get thirsty, cows would have to be milked—but that didn't happen. Some did go home, but they came back, and more people kept arriving.

"Two police cars arrived, and the superintendent from Ballinaboy got out, shouting, 'Who's the leader? Who's the person to talk to?' No one responded, so he tried again. 'Who's responsible for this?' And again, no one said anything. He wandered right into the middle of this crowd, and he said, 'I've got to talk to you people.'

"'I'm not a leader,' said Maureen McGrath, 'but I'm the wife of one of the jailed men. So, what do you have to say?'

"'Well,' he replied. 'I just want to say that I was sent here to see that everything is in order; no one is in danger or getting injured. I can see that you are policing yourselves very well, and I don't need to be here.'

"I often look back at that little incident, when thinking about how the relationship between people and the cops changed. I had never heard the expression 'policing yourselves.'"

Once the occupation of Shell's building site began, Betty and Fritz and their youngest son went every single day to support the protests. In the beginning, people in Ballinaboy felt that the refinery was their own problem, that those who lived a few miles away probably wouldn't be affected by Shell's project.

No one living right there ever moved away.

Frequently, meetings of those resisting the project were held in family kitchens, in the pub in Glenamoy, and later in Betty and Fritz's Kilcommon Hostel, at the Solidarity Camp, and the Inver community center.

"We were never very well organized," says Betty. "We did not have leaders—didn't want leaders—though of course there are always natural leaders, people who can speak better, while others, for one reason or another, choose to listen and participate in other ways."

BETTY AND I TALK ABOUT THE DIFFICULTIES of being involved in social movements, the time and energy required to work for a particular cause, and the possibilities that friends and family might not like what you're doing, or that the police might harass, beat, and arrest you, that judges will impose fines and send you to jail.

"But we were not unhappy doing it. There were unhappy moments, lots of anger, disagreement, sometimes a lot of hope, other times a lot of hopelessness as well.

"We weren't always being pestered by the police; we weren't always upset. We enjoyed being together, and we still want to do that, not on a large or organized scale, but just to be back in touch."

Shell's mercenaries did not directly threaten Betty or attack her family. Fritz wasn't arrested or beaten at protests, just pushed around a lot. Police poked around the hostel's car park, stopping and searching some visitors, demanding to know why they were going to Kilcommon Lodge Holiday Hostel.

At times, the police closed the road to facilitate Shell's vehicles traveling to the refinery's building site, preventing Betty from leaving home.

"But I live here," she'd remonstrate. "I need to go to the shop."

"No," they'd answer.

"Am I under house arrest?" she would ask. "Can you explain to me why I can't leave? Is there any order?"

"Our children were nervous," she remembers, "asking why we didn't put up cameras on the driveway. They wanted us to start locking our doors. After all, Shell's mercenaries were not under anyone's authority, no rules or regulations. I didn't think cameras would prevent someone doing whatever they pleased. A really nasty person could figure out how to avoid cameras."

I tell Betty about working with various peace and social justice groups in the United States over the years, about the death threats that I and others received. We talk about the growing fear in the United States that Big Brother is watching, reading citizens' mail, watching their email, listening to phone conversations.

"Yes, so, okay," she responds. "My big *crime* is that I didn't want a refinery in Ballinaboy and a pipeline in Glengad. Everyone can know that. If they read our emails, what can they really see? We communicated with a lot of different people, and we were always totally open about how we think about the cops, how we think about Shell.

"This is the opposite of paranoia," she avers. "I just refuse to be paranoid because if you start that, you will self-limit. Enormously. Unnecessarily. It's just not in my nature."

BETTY HAD BEEN THERE THE NIGHT MEN wearing balaclavas assaulted Willie Corduff. She'd gone home a few hours before it happened, and she still feels guilty for not having stayed. It was a dangerous situation, but she did not fear for herself.

Betty never contemplated dropping out of the movement, even though she was disillusioned when some people turned against one another. She struggled to overcome these feelings, but from working on other causes in Germany she knew that strife often happens when people get together to support or oppose something. People joined the Shell to Sea movement for different reasons, coming in with different political perspectives, different goals and ways to communicate. There were very young people and very old people, with various levels of ego investment. Most people had never done anything like this before. Still, the divisiveness came as quite a shock to Betty.

"What impressed me about all these difficulties," she says now, "was how they were most often overcome. How tolerant and trusting people were, how inclusive.

"Shell thought it would have a very quick catch: Build this thing before anyone notices. It's really quite unbelievable how unprofessional, with how little knowledge and how much arrogance, Shell acted. That was really the only big surprise—how unprofessional this multinational company turned out to be.

"Shell thought we wouldn't even know how to use a mobile phone."

Betty isn't sure how she managed to run a business and spend so much time opposing Shell's project. Visitors were understanding, offering a lot of solidarity. Each morning she rose at five o'clock, set the tables for breakfast, and went off to join the blockade from six to eight. Returning from this action, she made and served breakfast for her guests, tidied up the house, and then went straight back to the blockade from eleven until one p.m.

Resisters had different schedules, yet everyone managed to fit a chunk of time into their day. The police tried to prevent mothers like Betty from getting home in time to send their kids off to school. That didn't work.

"On one day of action when a little bit more was happening," Betty says, "there was more outside support at the site, so I told guests there might be arrests. When I came home they were listening to the local radio to see if anyone had been arrested, or whether they'd get their breakfast that morning.

"Some of our guests came along to protests, spent an hour there, talked to people and made donations. It's amazing when you're involved in something like this how many people you meet who say, 'Yes, at home we're fighting fracking or opposing nuclear power,' things like that.

"It was surprising to learn how many of our guests would connect with us. They knew what was going on in Erris, especially when the Rossport 5 were in jail."

SOME PEOPLE FIND IT HARD to believe that a campaign like this would be taking place in such an isolated part of the world: *Ireland has a long and tumultuous history, but it is now a peaceful country, its people determined to make visitors feel welcome and safe,* they think. *In Ireland, citizens cannot be hauled off to prison for trying to protect their environment. The police, unarmed and friendly, are not attacking ordinary people. Betty must be exaggerating.*

Police and local politicians tried to convince Betty to give up her opposition to Shell.

"'Come now,' they'd argue," she recalls. "'You have achieved a lot. It's so much safer around here. There's so much more awareness, more employment, and now Mayo County Council has more money.' They tried to praise us out of protesting."

Sometimes the police tried to stop her from returning home from a demonstration.

She plays out their part: "'Now, Mrs. Shultz,' they would say. 'You have a choice between protesting and looking after your children. One or the other. Not both.'

"They did things like that a lot to Willie Corduff and his family, trying to make them feel responsible for every 'not nice' thing that happened," she recalls. "When things were getting out of hand, the cops said to Mary Corduff, 'Well now, you've got what you wanted.'"

Norwegians who joined the Shell to Sea campaign found it difficult to understand the conflict between Irish police and ordinary people. In Norway, they told the Shultzes, the police learn that their place is with, not against, citizens.

"That's what *we* tried to say to the cops," Betty says. "'Look, the refinery will spoil *your* environment as well.' We asked them what might happen if gas or oil were found near where they live and their own mother went out to protect *their* environment. They'd have to confront their mother, who'd be trying to protect the environment in which her own children grew up.

"'You should not be *opposing* me,' we said to them. 'You should be standing *beside* me.'

"The Irish national police force," she points out, "was established by choosing men from local communities who were well liked, involved with Gaelic games, very integrated and very good people. This explains why people were so upset by the cops' behavior around here. The police acted unprofessionally. They were clumsy, escalating situations when they should have been calming things down."

Shell's public relations department promoted the idea that local businesses started thriving once the go-ahead was given to build the refinery. For a short while, a few people did make money. No one got rich. Moreover, this limited prosperity was short-lived. Betty made it clear that she wanted nothing to do with the refinery: Shell's employees could find lodging elsewhere.

"This is a tourist business," she says. "It has nothing to do with industrial development. We could not benefit from something that we opposed. Three or four businesses might have actually benefited from Shell's project. And many young people say they got valuable training from working with Shell. They have certificates now that might help them find work."

WHY ARE SOME PEOPLE WILLING to take great risks in order to oppose projects like Shell's refinery in County Mayo, while others believe that protests, blockades, and civil disobedience are futile? People are attacked, injured, sent to jail, and killed; pipelines spread across the land, mines are dug, forests clear-cut, refineries built. *Does resistance make any difference?*

Asked about this, the late Philip Berrigan, Second World War combat veteran, Catholic priest, and peace activist who spent eleven years of his life in prison for opposing war, said: "If you try to measure the difference you make, you will never make any difference."

Driving home from grocery shopping one afternoon, Betty passed a "SHELL OUT" sign, one of hundreds the campaign had spread across Mayo.

She reflects: "And I thought, okay, Shell's gone away. But it doesn't really matter very much to us. The refinery is still there. The pipeline is still there."

DESPITE HER OWN FAMILY'S PAIN and suffering, and their recognition that Shell never intended anything good for the people of Mayo, Betty feels sympathy for those who backed the company. "It matters more to the supporters and admirers of the company," she says, "because they feel let down by their big friend. They were loyal, and where's the loyalty from Shell now? I think there was more shock and disappointment from the people who believed in Shell. All the nice things that Shell was going to do for them will never happen.

"Shell's leaving wasn't that big of a surprise [to us]. It doesn't make a difference, really. A lot of boundaries were pushed around here. The police could see how far they might go without punishment or outcry, establishing this new way of dealing with protests. We often talked about the reality of what happened here possibly happening in many other places in Ireland as well.

"In the last two years, we've heard people say, 'Oh, we don't want another Rossport.' Or 'Remember what happened with the Corrib gas?' I think that communities and politicians don't want a mess like this to ever happen on their doorstep."

Colm and Gabrielle Henry, lifelong residents of Mayo

Musicians

There is no meaningful democracy in Ireland for communities. That is, if we mean democracy rule *by the people*. Ireland, a century after freeing itself from hundreds of years of colonization, has once again become a captured state. It has been re-colonized by multinational corporations.
—Dr. Rory Herne, the People's Forum, County Mayo, 2014

IN EARLY NOVEMBER 2014, OVER TEA and brown bread, then whiskey, I sit beside the Henrys' kitchen stove. I'm asked about my ancestry. In answer I quote my mother: "We come from a long line of horse thieves," a joke to express our family's lack of social status. No relatives on the *Mayflower*. Welsh, Irish, Scottish, English, a potpourri of DNA tossed into the melting pot. We never spoke about our roots. At some point, our distant relatives had arrived in North America, made their way to the Midwest, lived quiet lives, and died.

ONE OF SEVEN CHILDREN, TWO BOYS and five girls, Gabrielle

Henry grew up here in the village of Glengad, helping out on her parents' farm, walking to and from National School, playing games in the evening, romping on the beach.

"The village was a lovely place to live," she recalls. "Neighbors were more trusting and caring; they looked out for one another. Life had a nice easy pace. Now, after the struggle with Shell, it's much harder to trust anyone."

Colm, her husband, also grew up on a farm in County Mayo. His paternal aunts and uncles played instruments and sang. His mother's family, the Morans, were fine dancers and singers. Colm's mother gave Colm his first guitar when he was twelve. At fourteen, he was playing folk, country, and other forms of popular music with a band in dance halls and at carnivals.

Colm and Gabrielle's children loved music as well, and the family formed their own band, touring Ireland, England, and Scotland. Colm has performed in Nashville many times. Now his grandchildren are learning the value of music.

On both sides of Colm's family, men risked prison, exile, and death to free tenants from oppressive landlords, and to free Ireland from British occupation.

Anthony Henry, Colm's great-great uncle, was captured, imprisoned, and exiled to the United States. He never returned to Ireland. In one "strictly confidential" document now in Colm's possession and titled "Protection of Person and Property (Ireland) Act, 1881" and "Recommendation for Arrest," marked "Belmullet, County of Mayo," some authority figure writes that Anthony Henry was "vice chairman and chief organizer of the Geesla Branch of the Land League. No man in this part of the country was so ardent in promulgating and endeavoring to enforce Land League doctrine."

COLM AND GABRIELLE MET in the late '60s, married in 1971, and raised their five girls and three boys just down the road from her parents' farm. The Henrys' children went to the same schools their mother had attended. They played on the beach, went fishing, and swam in tidal pools as she had. Tourists stopped to admire the beautiful beach, says Gabrielle, "with its multicolored sands, heaving seas, crashing waves, and unspoiled wildlife. Summers were always glorious."

This was the old Ireland—families growing vegetables, raising sheep and chickens, giving what they could to those who had less than they did. No high-rise buildings, superhighways, or shopping centers. To those who thrive in fast-food urban worlds, it was a quaint but backward place.

In 2000, rumors began to swirl about the countryside, as the Henrys recall it. It appeared that natural gas was going to be brought in from wells far out at sea. Like most people in this region, the family knew little about gas refineries and pipelines. They assumed the gas would be processed at sea, and brought to land in a reliably safe manner, benefiting the Irish people and posing no danger to County Mayo's environment.

"Shell and its partners came like thieves in the night," remembers Gabrielle, "wanting to purchase properties to get the gas ashore, splitting the community, with some for and most opposed to the project."

One fine morning, Colm encountered men he did not know walking on the beach in front of his home. Obviously not on holiday, they seemed to be taking measurements of some kind.

In retrospect, the Henrys wonder, *What might have happened had we been able to foresee the tumultuous future? What might we, personally, have done to stop Shell Oil's project before this multinational began excavating our beloved beach, destroying the habitats of migratory sand martins and other wildlife?*

INEXPLICABLY, SHELL SET UP CAMERAS close to the Henrys' home, training high-intensity spotlights on the family's windows at night, watching their every movement. Men wearing black masks, dark gloves, and dark glasses lurked about, peering into the home with binoculars. Men holding cameras followed Colm and Gabrielle when they walked with their grandchildren on the beach. Workers erected high-security steel fencing to block access to the water, forcing the family to travel a long distance in order to enjoy picnics and swimming, things they had always taken for granted.

Colm and Gabrielle reported this harassment and intimidation to the police, who said their complaints were a civil matter; therefore, they could not offer help.

Years later, in a letter to the Planning Board's Oral Hearing into the Corrib Onshore Gas Pipeline Application, the Henrys would complain bitterly about surveillance in their community.

"For the last three years," they wrote, "our neighbors have also had to endure Big Brother–style fixed video cameras at different locations in our village, and the applicant [Shell] is proposing this would continue for at least the lifetime of the Corrib project, decades into the future."[1]

Trucks swollen with construction materials roared along the narrow road in front of the Henrys' home, forcing the family to find alternative ways to go shopping or to visit friends. Security guards intimidated the Henrys' adult daughters when they brought their children to visit their grandparents. This caused all of them huge anguish and distress, the effects of which they continue to feel years later. It would be compounded by the eventual placement of the dangerous pipeline within a few meters of their house.

Only after the media picked up the story and revealed the family's plight did the mercenaries move their cameras away from the Henrys' house; however, they left at the beach their high-security fencing and heavy squads of security personnel.

"The saddest part," say the Henrys, almost in unison, "is to know that our own national police were taking orders and direction from foreign mercenaries."

I TRY TO IMAGINE WAKING one morning to find bulldozers digging trenches in my own and my neighbors' yards, smashing our trees, polluting our air. We complain to the police, who say something about fine contractual print; about a legal entity called the "right of eminent domain." Materials vital to the national interest lie beneath our lawns, we are told: Keep quiet, obey the law, be grateful for the chance to sacrifice our children's health and welfare for the greater good.

AS WELL AS TRANSFORMING THE BEACH into a security zone, Shell ruins the Henrys' views of Broadhaven Bay, destroying swimming spots that form at low tide and disturbing marine life. Shell's boats drive away dolphins protected by international environmental and wildlife safeguarding agreements.

The Henrys write to the National Parks and Wildlife Service, part of the Department of the Environment, Heritage, and Local Government, to express their concern about damage to the nesting sites of sand martins. And they complain further: to the police, the Department of Justice, and the Department of Health and Children. The authorities are entirely uninterested. The Henrys bring their concerns about several matters to the attention of Shell's "community liaison" people, who promise to investigate. But they never get back in touch.

At one point, Shell holds an open question-and-answer session in Belmullet, but this turns out to be a charade. When people ask pointed questions, Shell's representatives demur, claiming that the employee who might clarify matters is unavailable at the moment.

"The assurances from Shell's experts seem to imply," write Colm and Gabrielle to the Planning Board, "that they have supreme powers over the physical laws of nature. This year's catastrophe in the Gulf of Mexico has shown how dangerous this attitude can be."[2]

Also in 2010, the (U.S.) National Wildlife Federation publishes a report titled "Assault on America: A Decade of Petroleum Company Disaster, Pollution, and Profit." Said the report:

> The BP Deepwater Horizon spill is truly a tragedy of our time. It should be used to take a closer and more comprehensive look at the full and continuing costs that the oil and gas industry continues to impose on society with its pollution, environmental degradation, habitat destruction, wildlife loss,

worker and community endangerment, health effects consequences, and loss of life....

Offshore: The U.S. Mineral Management Service (now Bureau of Ocean Energy Management, Regulation, and Enforcement) determined that 1,443 incidents occurred in the Outer Continental Shelf waters from 2001 to 2007. Of these incidents, 41 fatalities, 302 injuries, 476 fires, 356 pollution events were reported....

From 2000 to 2009, pipeline accidents accounted for 2,554 significant incidents, 161 fatalities, and 576 injuries in the United States....

Major oil spills are really only a small part of the real story. From 2000 to 2010, the oil and gas industry accounted for hundreds of deaths, explosions, fires, seeps, and spills as well as habitat and wildlife destruction in the United States. These disasters demonstrate a pattern of feeding America's addiction to oil, leaving in their wake sacrifice zones that affect communities, local economies, and other landscapes.[3]

DURING THE PLANNING BOARD'S HEARING, a gas pipeline accident in Iran kills several people, and a pipeline explodes in a suburb of San Francisco, to devastating effect.

"This [latest] incident," write the Henrys, "in particular highlights the folly of [Shell's] proposal, as we are told a rupture on the Corrib pipeline—also classified as 'suburban'—is impossible."

Oddly, the location of their own home does not appear on a 2008 impact statement prepared by Shell for a modified pipeline route, even though the route was intended to go right past their house.

A letter writer, Arthur Boland, tells the *Irish Times* that he finds it startling that a pod of dolphins has been herded out into the bay and sand martins are being prevented from nesting by netting. "May I ask exactly," asks Boland, "what terrorist blacklist dolphins and sand martins are on? What threat are they to Shell, whose website proclaims that it operates in 'environmentally and socially responsible ways'?"

One afternoon, Colm, his neighbor John Monaghan [son-in-law to Micheál Ó Seighin], and about twenty supporters walk to the Henrys' beach, intending to question Shell about authorization for continuing work on the pipeline. They have already written to the National Parks and Wildlife Service, part of the Department of the Environment, Heritage, and Local Government, to express their concern about damage to the nesting sites of sand martins. Thirteen of the people are arrested and released without charge. Later, Eamon Ryan, the Minister for Communications, Energy, and Natural Resources, issues an apology . . . for having failed to publish a work authorization on his website.

THE GAS BEING PIPED AT HIGH PRESSURE beneath the beach in front of the Henrys' home will be odorless, says Shell. No problem! And in the unlikely event that something does go wrong, people can call the company's local office!

"So, some farmer down the way lights a cigarette near a leak in the pipe, what am I to do?" scoffs Colm. "Call Shell to ask for help?"

At a meeting of the Planning Board on May 19, 2009, held in Broadhaven Bay Hotel in Belmullet, Irish Army Commandant Patrick Boyle, who'd served in Ireland and with the United Nations in Lebanon, has something to say. He warns there will be "horrific" consequences for people living around Rossport if Shell's pipeline "should rupture and explode. A hospital with a specialized burns unit and a fire station would have to be located in the area, due to the potential for fatalities and serious injury in a very isolated area."

Commandant Boyle goes on to say that there have been many pipeline accidents at pressure loads of 70 bar [approximately 1,015 pounds per square inch]—half the pressure Shell is proposing for the Corrib pipeline.

He cites as an example the July 2004 explosion in Ghislenghien, Belgium, in which twenty-four people had died and more than 120 were injured. Most of those killed were police and firefighters responding to reports of a gas leak in a pipe operated by Fluxys, a pipeline operator owned by Royal Dutch Shell.

WRITING TO THE PLANNING BOARD, on April 6, 2009, the Henrys express concern that aspects of Shell's pipeline are experimental:

> Nowhere in the world does a community have to live beside such an installation, and we do not accept the unknown risks that we would be forced to live with every day.... We have never been approached by anyone from Mayo County Council, Shell-Statoil-Marathon, Enterprise Oil, or anybody from the authorities to ask if we approve of the development, if we had worries or fears, or even if we had any questions. There has never been any proper public consultation on this project, and the only information put to the people has been advertising for Shell and their partners.

If the pipeline is built, the Henry family home will lose its value and, even if they want to move to safer ground they'll have to stay and accept the risks.

DRIVING ALONG WESTERN IRELAND'S narrow roads, visitors pass modest houses set upon small plots of land with a few sheep or cattle, peat smoke rising from chimneys. The people who live in these homes are soft-spoken, unassuming, easygoing and hospitable. Profit-driven corporations fail to grasp the meaning, the social reality of *place*. They do not understand, or care to know, that people who've lived on the same land, in the same community for generations feel a powerful sense of belonging there. These people might have political differences, but they share a love for the natural environment, a deep-rooted sense of *place*.

During the American War, as it is known in Southeast Asia, the Vietnamese were willing to endure intense bombing, vicious firefights, and exposure to Agent Orange and other toxic herbicides in order to remain where their families had been living for a thousand years. For centuries, invaders had tormented, tortured, and slaughtered the Vietnamese, but the people's attachment to their rivers and mountains, to jungles filled with tigers and elephants, bears and snakes, was unshakable. Powerful armies—from China, France, United States—failed to destroy the Vietnamese people's ancient sense of *place*.

Multinational corporations like Shell Oil might pay bribes to powerful politicians, use the police and hire mercenaries to attack their opponents, send honest men and women to prison, but they cannot obliterate the ancient love indigenous people feel for their land.

MIGHT THOSE WHO OPPOSED SHELL'S plans have taken their case to the European Union?

"You could," answers Colm. "But first you'd have to go through the district court, then the circuit court, the High Court and the Supreme Court before you'd get to Europe. And that would take years and years and years. Some people from here did go to the European Union, but nothing was done about it because Shell is just too powerful. They dictate to the government here."

And the police?

"We feel totally let down by them," says Gabrielle.

"If you go to them, it's like showing a cross to a vampire," adds Colm. "For a time, you couldn't even mention Shell on the radio. They take over everything. They do as they please.

"We've gone to court, and we've seen Gardai pick up the Bible and just lie and lie. At least they were contradicting each other," he laughs. "Somebody was obviously not telling the truth. Ah sure, they might be kind of nice to your face, but it's 'Sorry, we're just doing our job. That's all. Just doing our job.'

"When the guards first came here from other parts of the country," Colm remembers, "they didn't know what they were getting into, but there was lots of overtime and other perks. Then you'd run into a guard from Cork and ask him what he might be doing if something like this was coming into his community. 'I'd be doing the same as you're doing,' he would say, 'and I'd be doing it worse.'"

"It was the local ones who were the meanest," Gabrielle says. "That's the sad part now. Where there was trust at one time, now there's total mistrust. So now we wouldn't go to the guards, even if somebody broke into our house. No, we wouldn't ring these guys up.

"You see them on the street and they won't look you straight in the eye," she adds. "They might like to be friendly but they can't. And when Shell's security guards would take pictures of our children and we demanded they give them back, the mercenaries said, 'Oh, we don't have them anymore. The cameras weren't really turned on.'

"There was just no chance of getting justice. People might think we're exaggerating, but you can't exaggerate what happened in Mayo. You just can't."

AS A BOY, I WAS TAUGHT TO RESPECT the police. My father and other members of my family served in the Second World War. I believed the United States, the country in which we lived, was the freest, most powerful nation on earth. Like soldiers, members of the police represented honesty and courage, the willingness to risk and even to give their lives to protect their fellow Americans.

Later, living on the streets of New York City changed my view of the police. I felt helpless. They could beat, arrest, and even kill people with impunity. I grew to hate them.

Years later, I managed to step back from my hatred and ask more fundamental questions about the role police play in countries like the United States.

Who, really, are the police? They do not establish laws giving oil and gas companies the right to pollute and poison, to destroy our planet. They do not create public policies designed to protect the rich and punish the poor. The police don't live in luxurious homes paid for by hardworking taxpayers.

Police departments are paramilitary organizations, based upon strict hierarchies, group-think, and obedience to authority. Police officers are trained to follow orders given by superiors who take orders from their superiors who, all too often, take orders from corrupt politicians.

Police officers live and work in the context of worlds they did not create. Their job is to sustain the illusion that law and order are synonymous with justice. They are guardians of systems designed to hypnotize people into believing they live freely in free societies.

THEN THERE ARE MEN LIKE MAURICE McCABE, an Irish police officer who paid a terrible price for challenging dishonesty and corruption within An Garda Síochána. Although his story didn't intersect with the Shell resisters', there are parallels.

I was fortunate to meet Sergeant McCabe for coffee one afternoon, hoping to interview him for this book, but because of pending litigation he could not agree to that. I can, however, share part of his story.

"We have endured eight years of great suffering, private nightmare, public defamation, and state vilification," McCabe and his wife, Lorraine, told the *Irish Examiner* in February 2017, "arising solely from the determination of Maurice to ensure that the Garda Síochána adheres to decent and appropriate standards of policing in its dealings with the Irish people."[4]

The McCabes' personal lives and family life, including the lives of their five children, had been "systematically attacked" by state agencies. "We have also been the subject of a long and sustained campaign to destroy our characters in the eyes of the public and public representatives and in the eyes of the media," they said.

In a world where men and women follow orders, even ones that violate their own sacrosanct values, Sergeant McCabe obeyed his conscience, refusing to give up his campaign to create an honest, respectable Irish national police force. I knew about Maurice McCabe's efforts to expose wrongdoing within Ireland's police force, the relentless and extraordinarily vicious attacks on him and his family, and the actions of unscrupulous people who attempted, quite literally, to destroy him. Officer McCabe would most likely not wish to be called a hero. Nevertheless, it was my good fortune to have met this remarkable man.

"THE SAD SAGA OF THE CORRIB gas project has left our community in turmoil, and has caused untold suffering for many families, including our own," wrote Colm and Gabrielle to the Planning Board. "We have no confidence in the applicant [Shell Oil] being able to deliver a safe and environmentally sustainable project, the information provided has been sorely lacking in sufficient detail in vital areas of health and safety, and we have been left with the unaccountable legacy of poor decisions made by those far removed from the impacts of their decisions."

THREE YEARS AFTER I FIRST MET COLM and Gabrielle, we are standing outside of their home on a perfect blue-sky afternoon. The mercenaries are gone, the security fence running along the beach has been removed, and there's no physical sign of the struggle the Henrys waged to protect their family, their home, and their environment from an imperious company.

"Once in a while," says Colm, "an American couple would stop by our house. 'What a beautiful place,' they say. 'Would you take a blank check for it?' Well, they wouldn't say that now that there's a gas pipeline running right outside our front door.

"Shell wanted to show us," he continues, "to defeat us. But they didn't. We've made such great friends, people who are close to us. Shell's effect on the community is still there, and it will be; the wind won't blow it away. We met a doctor who'd worked for Shell, and he apologized for that, telling us that he admired the way we handled ourselves. We talked to several guards [from other places], in their civilian clothes, who said that if this kind of thing happened in their community they'd be doing the same thing as us. Not one local guard did come to us to say they were sorry. It'd be different, less hurtful, if we didn't personally know them.

"There were thousands of stones down at the beach that we

could have used against the guards, but we didn't. Not once. We were peaceful. Always. We don't mind talking about it, even after some years now, but we will never, ever accept it. Never. 1.7 million euros' worth of gas flowing through that pipe every day, and not one penny coming back to Ireland."

In other words, you're not getting free gas from Irish waters?

Colm and Gabrielle laugh without humor. "Our town didn't even get a gas supply," answers Gabrielle.

According to a report by Andy Storey and Michael McCaughan for Afri [Action from Ireland]:

> An international study in 2002 found that only Cameroon took a lower share of the revenues from its own oil or gas resources than Ireland. . . . Ireland, by contrast, demands no state shareholding in any resource finds, nor does it demand royalty payments. . . . The Corrib gas field will probably be half depleted before any tax is paid at all. . . . We are simply giving away our gas—at a time when the state's finances are severely strapped. . . . A recent wave of resource nationalism has seen governments around the world take back control of natural resources that were previously in the hands of foreign companies. . . . In none of the cases did the affected companies (including Shell) walk away from their investments, despite their dissatisfaction with the new regimes.[5]

DID THE HENRYS AND THEIR FRIENDS *ever feel like giving up?*

No, they did not. People were still able to laugh. Humor, one of the things that has sustained the Irish throughout history, helped keep the resistance going. It was at times hard to sleep at night, knowing what they were up against, fighting to save their environment and their heritage, aware that people and institutions they'd counted on for support had abandoned them.

"We don't blame Shell in a way, because this is what they do all over the world," Colm says. "It's the government we blame. When you become a politician, you have to become a professional liar.

"Ireland is not the country we grew up in," he says. "There's no respect."

Can you change it?

"Revolution," says Gabrielle quietly, with the subdued hope of someone who has known the look of a better world.

Liamy and Geradine MacNally

Journalist

> I hear the plaintive cry of the Ogoni plains mourning the birds that no longer sing at dawn; I hear the dirge for trees whose branches wither in the blaze of gas flares, whose roots lie in infertile graves. The brimming streams gurgle no more, their harvest floats on waters poisoned by oil spillages.
> —Ken Saro-Wiwa, Nigerian writer, social activist, Nobel Peace Prize nominee, and president of the Movement for the Survival of the Ogoni People, in a statement made on Nigerian national television, shortly before his execution on false charges of murder

LIAMY MACNALLY IS A PROFESSIONAL JOURNALIST who believes in and practices truth telling. A lifelong resident of Westport, County Mayo, he is working full-time for Midwest Radio when gas is first discovered off the northwest coast of Mayo in 1996. Over the next few years, he listens to the hype—grandiose promises of riches, jobs, an exciting new beginning for Mayo. In 2002, he learns that Enterprise Oil, soon to be sold to Royal

Dutch Shell, intends to build an experimental gas refinery outside of a small town not far from his family home. Liamy writes about Shell's plans, invites guest speakers pro and con to appear on news programs, and discovers the many ways that money corrupts ordinary, otherwise fine people.

On a beautiful September day in 2017, Liamy welcomes Terence Conway and me into his home. His wife, Ger, sets out tea and cookies, then leaves to spend time in her garden. Liamy begins our conversation with a story about Brian O'Cathain, managing director for Ireland of Enterprise Oil, the company that discovered gas long before Shell arrived.

"I am at a Mayo County Council meeting when Brian comes to talk," remembers Liamy, "and he's nervous as a kitten because he thinks he's going to get a dressing down. The councilors, he assumes, will want the local people to know what is really going on, what is at stake, and so on.

"Instead, councilors are all coming to him, cap in hand. 'Mr. O'Cathain,' they say, 'will you please put gas into my village?' "

To Liamy, this is "an embarrassing meeting. No one asks Brian any hard questions, and he steps out with a big Cheshire cat grin. He is a good guy to talk to, a very affable man, and walking away from there he says, 'God, I thought I was going to get chewed, and nothing happened.' He can hardly believe his luck.

"You know, we have a term in Ireland—the 'Gombeen man.' This was the person who did everything for the English-appointed landlord, or the landlord's agent, and exploited the destitute during the Great Hunger—the 'water boy' who would do absolutely anything for the English."

Shell takes over from Enterprise, and immediately appoints public relations people to promote its refinery. Committed to giving everyone the chance to talk on air, regardless of whether he agrees with their views, Liamy invites supporters of and objectors to Shell's proposed refinery. He also writes a weekly opinion piece for the local newspaper, *Mayo News*.

"I'd write a column, exploring issues, and I would be getting a bit of flack, especially from Shell," he says. "Sometimes the editor wasn't too happy about that."

Liamy notices that at press conferences someone from Shell is always "sitting in his pocket"—getting too close to him. Everywhere he goes for anything to do with the gas project, people warn him he is being targeted by Shell.

"And I was. I am always their target because radio is a powerful medium. More people listen to the radio than read the papers."

So you had a following?

"Oh yes, that's a nice way of putting it," Liam laughs. "People knew that I would be covering the Shell story. National radio would pick up on some stories, and I'd send things off to the *Irish Times*."

Liamy often drives the 80 km, or about an hour and a quarter drive, northwest from Westport to Belmullet for meetings and to interview people. Many Irish roads are well kept and there are stretches of four-lane highways, but even short distances can feel as though you are always circling back or, somehow, twisting in the wrong direction.

"We're here in eastern Mayo," says Liamy, "and it's difficult getting over to Belmullet in the northwest. It's a long drive and an expensive thing to do for a local radio station. We didn't have the personnel, we had to [hire] a car in order to get a story. I'll give credit to the [station's] owners, because all this cost a lot of money.

"We discovered that something happened over in the Shetland Islands [part of the UK, north of Scotland], going back to the 'sixties. All of the big oil companies came in, and it looked like the islands were going to be swamped. But they had what we would call a county manager, not elected, a council official, and he said, 'If we don't do something about the Shetland Islands, we will be dictated to by the oil companies.'

"So he and his councilors do a lot of knocking on the doors of the national politicians, visiting the houses of Parliament in London," continues Liamy. "Eventually they draw up the Zetland [Shetland] Islands Act, giving huge powers to the local council, so the council can dictate who brings in oil or gas. You see, Shell is buying one section of an island, Exxon another, Texaco another. They are buying up parts of the islands so they can build refineries.

"The council calls all the oil companies and tells them, 'This is where we'll put the refinery. We own the land, and the oil companies will pay us rent. Build the thing, but we're going to have one refinery, not ten.' The oil and gas companies build the biggest refinery in Europe, called Sullom Voe Terminal, and every ship bringing oil or gas into it has to pay a tariff to the council."

Liamy shares this story with the owner of the radio station, who encourages him to visit Scotland and find out more.

Liamy and a few other journalists will go over for several days, and Liamy will send back reports to the station each day. "Now, before I go to Scotland, I write to a Director of Services in Mayo County Council, inviting him to come along. This is the council official who'll have many dealings with Shell. But he refuses to join us."

Returning from Scotland, Liamy offers to address a Mayo County Council meeting, where he'll explain how the Shetland Islands created an agreement with multinational oil companies. Mayo County Council does not invite this well-respected reporter to speak, so Liamy produces a radio program about his trip.

"I find out there were only two fiddlers left on the Shetland Islands before the discovery of oil and gas," he reports. "Following the Act and the accrual of levies, local politicians set up a music school. Today they have musicians available to all of the island's schools. And now there are loads of fiddle players. They set up schemes for farmers; one man who wants to go into pedigree cattle is supported. He receives funding to help him do that. The fishermen get funding for new boats and equipment. The older people want to live in their own homes as long as they can, so the [authorities] create home-care programs for them to stay at home.

"All of these benefits, the whole social strata, derive from the money that comes from the oil and gas. And I said, 'Why don't we do something like this so the Mayo people can actually benefit from Shell's gas?'"

He seems incredulous when he says, "It never happened here. [There were] never any benefits. When people say, 'Oh, but we got gas in Westport, we got gas in Castlebar,' the reason for that is the Rossport 5. It was their jailing that allowed concessions from Shell and the government. Otherwise, there'd have been nothing.

"The bottom line is that the government sided with Shell. Shell always dictated the terms and the government replied, 'Yes sir, no sir, three bags full, sir.' That's just the way it was.

"The Shell thing split the community. The company bought their way into certain people."

By way of example, Liamy tells this story: "I got a call one day from one of the senior guys at the radio station who said I should lay off the Shell story. 'And why should I do that?' I asked. 'Well now, Liamy,' he says, 'the Shell story involves the company, it involves the [Mayo County] Council, it involves the government, it involves politicians, it involves local people. And you know, you're giving some local people too much time on the airwaves. You know Shell is spending a lot of money up in north Mayo.'

"'Now hold on,' I say. 'Who has made a complaint about me?' And he says, 'One of our advertisers.' Shell felt it wasn't [being made] welcome [and that] I was giving too much time on the radio to objectors. And I said, 'Listen, if people have a concern about anything, I'll put a microphone under them, and if your concern is legit, then there'll be a follow-through. If not, it disappears.'"

LIAMY COVERED MANY DEMONSTRATIONS against Shell, including a blockade of the entrance to the refinery soon after the Rossport 5 were sent to prison.

The protesters were always peaceful, he says. It was "like a football game—no trouble at all. Protesters just weren't letting trucks in or out of the worksite.

"The police arrived and a young sergeant, Mick Murray, stepped forward. 'I'm a cop, an Irish Garda, like your own son or daughter. I come from a rural community. You have a right to protest. But I'm a guardian of the peace. I have my job to do, and if anything happens I will have to take action. If people are peaceful, I will have no problem.'

"There were families there, children," Liamy says. "I'd say that man's speech was one of the best I'd ever heard in my life, because he was being honest. There wasn't a scintilla of trouble."

It wasn't always the case, however. "Things happened later on, when cops brought in the heavy gang from Dublin—a riot squad, some just itching for trouble," Liamy relates. "I have seen photographs of these guys with batons, going around antagonizing people. You can see it in film footage, cops who wanted action in this place. One particular riot squad guy from Dublin was a thug—arrogant, an absolute disgrace to decent guards.

"I mean," he says, "these are rural people from our county, who are standing out to protest. And they are met by the full force of the state. Our force is different than yours in America. We don't send cops out with guns, but we send them out with bloody attitudes. It was scandalous what happened to people like Pat O'Donnell, Willie Corduff, and others. Absolutely scandalous. Don't forget: These people are farmers, fishermen, teachers. That's who they are, what they are, protesting something they know is not right for where they live."

In the earliest days of the resistance, the protesters were almost all local. "Later on," says Liamy, "other people did come in, some types you might call 'crusties,' who follow environmental causes. Some were good, peaceful. Others liked a bit of aggro."

Liamy is a straightforward, honorable journalist who practices journalism ethically, with a mind to fairness as well as to getting the full story. "I would sum it up," he states plainly, "by saying that the people of Erris were betrayed by the government, by Mayo County Council, and by the oil company. Betrayed. Completely. I have no qualm at all about saying that. We went to various meetings and court hearings, and we heard all these lies being told at various levels, even perjury in court. I witnessed it.

"I felt pressure early on because I knew that some people were giving out. Shell would always want to counteract any story, and I always gave them the right of reply. But then they would try to isolate me. They'd send in complaints to the station and I'd say, 'Well, let them write to the Broadcasting Complaints Commission. If they have a problem with my broadcasting, let them make a formal complaint. If they have a legal problem with something I do, let them sue me.'"

Liamy spoke and wrote about Shell's crimes against the Ogoni people in Nigeria. The only way to get through to companies like this, he concluded, is to stop buying their products.

"It won't make any difference, people tell me," he says. "'Well maybe not,' I say. 'But it will make a difference to me.'"

MacNally tells listeners to his radio program about the Ogonis' suffering, the destruction of their environment, Shell's collusion with Nigeria's military dictatorship and refusal to intervene on behalf of Ken Saro-Wiwa, and the illicit framing and execution of Saro-Wiwa and eight fellow environmentalists by that junta.

"Any mention of the Ogoni, or drilling in the Arctic, things like that, would drive Shell ballistic," Liamy tells me. "I've always said if you want to know what this thing in Mayo is all about, go back to the beginning. Shell wanted to bring in a gas pipe with 340-bar [almost 5,000 pounds per square inch] pressure close to people's homes. Now, what does that mean? Nothing to most people. If you break it down, the pressure that comes into a private house from the street is something like 2.5 bar [about 36 psi]. As Willie Corduff said, 'We have the choice of being incinerated or drowned in a tsunami. How do you want to die?'

"I'm most disappointed with our own county councilors. They're the bottom feeders politically. In saying that, there were a few great people among them, but Shell's opponents were dismissed. [And I'm] disappointed with the police. I knew some of the Gardai, some of the senior guards, and certainly expected fair play. They'll say their hands were tied, they were just taking orders from the higher-ups. But what does it take for a man to have balls? What does it take for someone to stand up and say, 'Hold on. Hold on, what's happening here is not right.'

"It's always about *Follow the money*. It will all come out some day, when some person will sniff his or her nose at this story and follow the money. I'll guarantee this, [wager] my house on it. A few days after the initial planning permission for the gas refinery at Ballinaboy was refused, Shell's people were walking into the Taoiseach's [prime minister's] office. I mean, *please*. Talk about a banana republic. It's just crazy carrying on.

"We have a Freedom of Information Act in Ireland, but it costs 15 euros to apply and could cost several thousand euros to [actually] get the information. Who can afford that? There's a lot of poison, a lot of money to follow."

AS HE CONTINUED TO FILE STORIES, subtle pressure mounted on Liamy. He was called in for several meetings with the radio station boss, who decided to take him off the gas story. He was not allowed to report on it, even though he had won awards for his reports and coverage of the Corrib gas field.

"The station was letting someone go from the newsroom," he says, "and I was the FOC, Father of the Chapel, or union leader. So, they asked my three female colleagues and myself which one of us would like to leave, and no one wanted to. Well, I said, if I were to go I'd need a package. We couldn't agree on terms. Then they said I would *have* to leave. I had no choice then, and was made redundant after almost fourteen years of service. I took my case to the Employment Appeals Tribunal and won three years later."

He laughs a little, ruefully. "This is the crazy thing in Irish law," he continues. "The ruling is: 'Okay, you were unfairly dismissed, and if you'd been working since then you would have earned so much. But in fact, you *have* been working for the last year and a half, so we'll deduct that from your award.' So, because I got work, that's credited to the employer who actually lost the case."

Liamy laughs again. "Well, now, we have a saying: 'A happy man is one who enjoys the scenery on a detour.' I was the only male there, so I guess I could have claimed sex discrimination."

JOURNALIST

WHAT MIGHT LIAMY SAY TO A YOUNG *aspiring journalist, like the college students I taught for many years who were often stymied by the fear of the consequences of writing or speaking about controversial subjects?*

"They're fearful," he replies, "because they try to map out their life ahead of them. But you can't do that. You cannot map your life. You know, we always say, 'Take care,' but we should change that to 'Take risks.'

"If a young aspiring journalist asks me about taking risks, I'll ask them what kind of Ireland they hope to see for their children. Do they want an Ireland where people are afraid to take risks? Life is not *about* risks. Life *is* a risk. *Growth* is risk. You're not going to grow by not taking risks.

"I write an article for the local paper every two weeks, whether it's challenging a council, the government, the Church, or something about Europe.

"I'd like to talk about the kind of Ireland I want, not the kind the European Union wants. Our government is brilliant at bullying people, frightening people. I don't have kids, but if I did they'd be paying for generations for the actions of the same people who brought Shell to Mayo, causing mayhem and pain, subjecting people to Gardai beatings and stints in jail, and damaging communities. We need to raise a generation who can follow truth and ask hard questions. Truth always travels a narrow path . . . and can be lonesome."

Maura Harrington (left) with peace and social justice activist, Margaretta D'Arcy

Teacher

> What is known is that the Irish people have no right whatever to their own natural resources as a result of that deal [secret deal between gas developers and Irish politicians].[1]
> —MIRIAM COTTON, IRISH JOURNALIST, MEDIA ACTIVIST, AND COFOUNDER OF THE MEDIA MONITORING SITE MEDIABITE.ORG

"IT WAS ALWAYS JUST SUFFICIENT," says Maura Harrington, "to know that this beauty is there. That was what got me first, the fact that an oil company would decide to come into this place. And then it took longer to dawn on me that a multinational corporation would not only be allowed to do whatever it wanted here, but that the government would actually encourage and facilitate Shell Oil's plans."

In Terence Conway's videos of the uprising against Shell, Maura sits in the bucket of an earth-mover, attempting to block Shell's movement along a privately owned road. In another video, Gardai smash Maura's windshield, drag her out of her car. In another, she stands in the rain at the gate of the refinery construction site, facing police who charge into protesters, beating youth, women, elders, and the infirm.

"I did get mad," says Maura. "And I certainly did not employ academic tactics just to calm down. We, people who have always lived here, learned so much about gas pipes, chemicals, and refineries. And we wanted to impart all of it to anybody who'd listen.

"The local bishop at the time was clueless, the local parish priest acted as chair of this pro-gas group, One Voice for Erris. There was absolutely no critical analysis. The oil company was promising a new swimming pool for the local town of Belmullet, and lots of other things like fiber optics. People believed they were going to get this, they might get that.

"At this stage, no one had the faintest idea what was going on. We hadn't a clue. I did go to a fair few of the One Voice for Erris meetings. They were acrimonious. But this group died with a whimper one year on."

MAURA RECEIVES A CALL FROM SISTER Majella McCarron, who wants to talk about Shell's new project. Sister Majella had grown up in rural County Fermanagh, joined the Missionary Institute of Our Lady of Apostles in Cork in 1956, and spent thirty years in Nigeria, teaching in a secondary school and lecturing in education at the University of Lagos.

A close friend of Ken Saro-Wiwa, founder of the Movement for the Survival of the Ogoni People (MOSOP), Sister Majella witnessed the crimes Shell Oil committed against indigenous Nigerians.

She shares with Maura her concerns about the future of Erris, County Mayo.

Maura listens, but thinks things will be different in Ireland from what Sister Majella describes in Nigeria. Nearing fifty, Maura believes that even though Ireland might not be a perfect democracy, public officials are still willing to listen to careful, well-researched arguments about vital issues.

"Today," she laughs, "I'd consider anyone who would say something like that to be hopelessly naïve!"

"People said, 'Have a nice quiet life for yourself, Maura.' But I had come out of that comfort box a long time ago. I will never forget coming across the *precautionary principle*—that is, whether the proposed environment [Shell's County Mayo project site] might pose any untoward level of danger.

"And I really thought this would be a silver bullet. It will kill Shell's plans stone dead. It isn't going to happen because all I have to do is research [around] this precautionary principle, apply it to the company's proposal, and this will be seen and read by the local planning authority. That will be it: justice done.

"I don't claim to represent anybody. I do what I do. I've always had an instinctive aversion to doing things I did not like, and thus I have never spoken to anyone from Shell Oil. Shell to Sea never met anyone behind closed doors. I did my research, made submissions, but it didn't make any difference. Working within the system doesn't pay, because being 'nice' gets you nowhere."

> FREEDOM OF PEACEFUL ASSEMBLY. Subject to public order and morality, the right of citizens to peaceful assembly, without arms, is guaranteed by Article 40.6.1.
> — DIRECTIVE, PRINCIPLES OF SOCIAL POLICY, CONSTITUTION OF THE REPUBLIC OF IRELAND

OPPONENTS OF THE REFINERY were shocked to discover that their fight wasn't with Shell alone, but also with their own government. Maura and fellow teachers like Niall King and Micheál Ó Seighin, civil servants all their adult lives, believe that common sense will inevitably prevail. With enough hard work, sound research, and careful reason, they are sure, people will listen to their concerns about Shell's dangerous plans.

But it was not to be so. "It was an eye opener," Maura says. "We did not abandon the state; the state declared war on us. Our Irish Constitution says: 'Fidelity to the nation, and loyalty to the state.'

"These are the duties of all citizens. And it's not like you're walking around singing, 'I'm being faithful to the state.' It would just be part of our makeup, what we learn in school, from our family, and from the Church."

DISTRICT COURT, BELMULLET, COUNTY Mayo. September 11, 2013: Maura Harrington, a diminutive woman wearing large round glasses, does not need a microphone to be heard. She addresses the district court with the confidence of a skilled schoolteacher.

The police have attacked Maura, tossing her into ditches, knocking her unconscious, and carting her off to prison. Once, they accused her of slapping a security guard, a peculiar charge given that she must have been standing on a ladder when she lashed out at one of those tall men.

Maura is here in court to talk about police misconduct in County Mayo during the years she and others have been struggling to halt construction of Shell's project. She intends to address the latest charges against her, under Sections 8 and 9 of the Public Order Act and Section 53 of the Road Traffic Act, alleged to have occurred on August 1 and 3, 2012.

Maura asks that a *nolle prosequi* (voluntary dismissal by the prosecutor of the charges against her) be entered in this instance or, failing this, that the hearing of the allegations against her be deferred "until such time as an independent, international panel commission has issued a report on the policing of the Shell/Corrib project with specific reference to practices, policies, procedures, human rights, bribery, corruption and collusion."[2]

Maura tells the court that since 2007 a number of human rights organizations investigating the situation in County Mayo have "expressed concern about the quality of policing affecting the protection of the right to protest."

These organizations, she explains, work within the framework of the UN Declaration on Human Rights Defenders, adopted in 1998 on the fiftieth anniversary of the United Nations' Universal Declaration of Human Rights.

From February 23 to February 27, 2007, an International Fact-Finding Delegation from Global Community Monitor (GCM)

USA had visited County Mayo, after which this organization published a report[2] highly critical of the way police treat those who demonstrate against Shell. The report states that police in Mayo were "not trained in the necessary skills needed to manage peaceful demonstrations and civil disobedience actions, maintain order, and protect the rights of free speech," and that they are "not utilizing training if they have [even] had it."[3]

"There is evidence," the report continues, "of excessive physical force by Gardai against peaceful protesters who were prepared to be arrested, which resulted in serious injury. There is evidence from videos of youth, women, and the elderly being pushed and beaten by Gardai without provocation. Even high-ranking officers were personally involved in beating up protesters."

Furthermore, as Maura points out, citing the report, police did not display their identifications—badges or name tags—at protests; nor did they warn anyone before taking physical measures against the peaceful resisters.

"Local and national authorities of the Irish government," stated these observers, "must immediately recognize that the situation in County Mayo could result in further serious injury to protesters, the public, and members of the Gardai. Action should be taken to restore order and peace to the region through the intervention of neutral third parties."

MAURA HAS SERVED A TWO-YEAR DRIVING ban together with a total of fifty days in the Dochas Centre, Mountjoy Prison—including a twenty-eight-day sentence for participating in a community effort on June 11, 2007, to keep Shell from using a private road to move machinery to the sea. She served fourteen days for nonpayment of a €1,000 contribution to the Garda Benevolent Fund arising from that conviction together with other fines totaling €2,700.

"Shell to Sea started as a local campaign," she says, "but as it went on we couldn't believe our ears about the actual terms and conditions of the pipeline and refinery, and finally it dawned on us just how bad the whole thing was."

At her trial for assaulting a police officer, Maura pointed out that the police were videotaping the demonstration where this violence supposedly took place. The officer who accused her of slapping him said that he felt humiliated.

"And Judge Mary Devins went into her spiel about my being a teacher and servant of the state. They don't like that. If you're on the dole or something like that, it's different. I don't know if she expected me to appeal, but she fined me a thousand euros for the 'assault,' and another thousand to be paid to the Garda Benevolent Fund.

"The irony is that Mary Devins used to be quite stern when it came to police misbehavior, coming down on cops like a ton of bricks—up until she started getting our resisters in Shell to Sea before her. I remember one time when a cop wanted to hand a summons to a man, and so he followed him into a nightclub and tried to shove it down this man's trouser pocket. Judge Devins lit into him; she had plenty to say.

"But when Shell to Sea came before her, we were like a red flag to a bull. I mean, she lost the class [her objectivity] entirely.

"We asked Judge Devins to recuse herself from hearing Shell to Sea cases because of the perception of bias. She was the wife of a government minister who supported Shell's project, and her judgments were political and media-driven, not based on law.

"Then she sentenced me to one month in jail. I mean, she should have had the wit to tell me to contribute one thousand euros to the children's hospital, or something child-related. After all, I will always be a teacher.

"What smarts is that after she sentenced me, and after whatever tirade she put off at that stage, a kind of byline for the media, she directed that I should undergo psychiatric assessment. She'd made a big statement in the beginning that she didn't want to hear anything about sociology or ecology or politics."

As the police drove Maura to Mountjoy Prison, she listened to news accounts of the judge's assertion that she needed psychiatric care. The following day, the judge's scurrilous comments appeared in many Irish newspapers.

"There might be times when psychiatric reports are needed in court, but that's *before* sentencing. So, she resorts to the age-old female thing of 'witches.' You know, tie the accused to a chair, then dump her into water to see if she will float or sink. In a patriarchal world, strong women who have their own views, and do whatever they wish to do, are still looked down upon.

"The 'Mad Maura' label did make some of Shell's opponents uncomfortable. Respectable people can be intimidated by this kind of portrayal of a resister. But once you keep doing what you set out to do, and do not waver, people will conclude you can't be all that crazy.

"I remain forever grateful to Senator David Norris," says Maura, "who asked in the Seanad [Irish Senate] on March 12, 2009: 'Are we returning to Eastern Europe? Is [Judge Devins's order] an attempt to use psychiatry to control political expression?'

"It's funny because now every time anyone from Shell to Sea is up to the court, the defendants make huge speeches about all these things, especially ecology."

MAURA TELLS THE COURT ON SEPTEMBER 11, 2013, "I have been before this and other courts on many occasions since 2009 on exclusively Shell/Corrib-related cases. I have accepted the verdicts handed down at District Court or, on appeal, at Circuit Court level.

"While I continue to accept the bona fides of the majority of officers of the court," she continues, "I can no longer accept the *bona fides* of the police who are present as officers of the court today or, indeed, of any police with a connection to Shell/Corrib, up to and including the Commissioner.

"I use the term 'police' deliberately," Maura testifies, "because I believe the term Garda Síochána is properly reserved for those members of the force who remember the oath they took . . . and who act in accordance with all regulations of AGS [An Garda Síochána, the Irish police force]. The mission statement of AGS, 'Working with Communities to Protect and Serve,' was torn up in Erris in the early hours of October 3, 2006, when I believe a subverted state, acting contrary to the common good, turned its police force on its own people at the behest of Shell.

"For some time, I have chosen to take the Declaration in court since, on too many occasions I witnessed police officers of the court mumble the oath over the Bible, then cast it aside and proceed to commit perjury—with impunity.

"I accept this is the 'appalling vista' for the presiding judge, but I can no longer place myself in the legal jeopardy that is brought about by political policing where perjuring police do more to dishonor the institution of the court—and the rule of law—than any so-called subversive group could hope to do.

"I have done, and will continue, to put myself 'on the frontline' in defense of Place but cannot and will not place myself in legal jeopardy.

"In this instance, justice delayed is justice—and law—affirmed."

IN HER TWELVE-PAGE STATEMENT, Maura talks about independent reports that support her assertions.

Frontline Defenders, an international foundation for the protection of human rights defenders, commissioned an independent report of the Shell/Corrib situation, culminating in the publication in 2010 of "Breakdown in Trust: A Report on the Corrib Gas Dispute" by Brian Barrington. This report recommended that An Garda Síochána appoint "a trained lawyer with relevant experience as human rights adviser...that that adviser not only review police policies and practice generally, but also provide input into the planning of operations. Specifically, in the context of the Corrib dispute, it would be helpful to provide guidance to Garda Síochána on how they can best discharge their functions in an impartial way."[4]

In November 2012, ten community members, including seven from Shell to Sea, met with UN special rapporteur Margaret

Sekaggya to talk about police violence in Mayo, the "behavior of private [Shell-hired] security, the democratic deficit in the planning process, surveillance and harassment, selectivity in the application of the law, the undermining and stigmatization of campaigners by the judiciary, the politicization of the judicial process, and the ineffectiveness of designated oversight bodies, in particular the Garda Ombudsman [Commission]."[5]

In March 2013 Margaret Sekaggya presented her report, in which she speaks of credible reports and evidence including video footage, indicating the existence of a pattern of intimidation, harassment, surveillance, and criminalization of those peacefully opposing the Corrib Gas project. She also mentions reports about the actions of the private security firm I-RSM, which Shell employs.

Ms. Sekaggya called on the Irish government to "investigate all allegation[s] and reports of intimidation, harassment, and surveillance in the context of the Corrib Gas dispute in a prompt and impartial manner."

"I participated in all these processes," testifies Maura, "and come before this court today numbered among those acknowledged by Global Community Monitor, Frontline Defenders, and the UN special rapporteur as a human rights defender within the framework of the Universal Declaration of Human Rights."

> The state is bound to protect the personal rights of the citizen,
> and in particular to defend the life, person, good name,
> and property rights of every citizen.
> —ARTICLE 40.3, CONSTITUTION OF THE
> REPUBLIC OF IRELAND

MAURA IS NOTHING IF NOT OBSERVANT of human nature, I have learned. "When one side isn't reasonable to begin with," she tells me, "you cannot have a reasonable conversation. We live on the periphery of Europe, we are white Europeans generally speaking, and we live in a country that is fairly well perceived as being a working democracy. It's much easier for people to believe what's happening to the Ogoni in Nigeria. And of course, what did happen to them, in relative terms, is off the scale compared to what happened here.

"But then we are just at the initial stage. If Shell is able to get away with this, and the state allows it, they will be as careless here as they were in Nigeria. It's easier for even liberal-educated people to imagine Shell doing something terrible in a far-off place like Nigeria, than to think that it could happen in Ireland. I think that is a big difficulty for us. It took actual experience to disabuse us of the notion *where* we were living and *what* we were living meant anything.

"Corporations are psychopaths. They need to be demolished. It's hard for most people to imagine how cruel, vicious, and deadly a corporation like Shell really is. They produce such a delightful veneer, with a whole stable of clean, scrubbed-up men and women in fashionable clothing working night and day to keep people from seeing the rottenness, pillage, robbery, and murder behind this façade."

Maura points out, "We have a long history in Ireland of creating tales and storytelling. That's always been a fantastic part of education, and it enabled us to spot Shell's spin-masters from miles away. [Because] they were not good at storytelling, they didn't fool us that way at all.

"In springtime '08, Shell started a psy-ops operation in Mayo, with the *Solitaire*, Shell's pipe-laying ship, coming back to Broadhaven Bay. There was all of this hyper buildup. The *Solitaire* was twice the size of Croke Park [sports arena] in Dublin—one of the largest ships on earth. And then it docked at Kelly Begs, Donegal, the county just north of Mayo. It was there for some thirty-two months.

"All this was deliberately designed to make Shell's opponents feel powerless.

"At some stage, I'm not sure when, but the thought came on big—you know, it appeared to me, wherever it might have come from, I did have to consider it: I mean, you can't physically attack a massive thing like Shell's ship. So, I began a hunger strike."

THE PRACTICE OF THE HUNGER STRIKE originated in medieval Ireland where, as *troscadh* (fasting against a person) or *cealachan* (achieving justice by starvation), it had a place in the civil code, the *Senchus Mor*. David Beresford, in his compelling 1987 book *Ten Dead Men: The Story of the 1981 Irish Hunger Strike*, explains.

> The code specified the circumstances in which it could be used to recover a debt, or right a perceived injustice, the complainant fasting on the doorstep of the defendant. If the hunger striker was allowed to die the person at whose door he starved himself was held responsible for his death and had to pay compensation to his family. It is probable that such fasting had particular moral force at the time because of the honor attached to hospitality and the dishonor of having a person starving outside one's house.[6]

Irish history is replete with stories of hunger strikers attempting to use the practice as a means of achieving justice, often dying as a result. Beresford's book includes many examples, including these:

> During the 1917 Irish War of Independence, "when prisoners at Dublin's Mountjoy jail refused to eat after being brutally treated for refusing to wear prison clothes and to do prison work, their leader, thirty-two-year-old Thomas Ashe, died after being force-fed. An estimated 30,000 to 40,000 mourners attended his funeral. . . .
>
> Terence MacSwiney, Lord Mayor of Cork, forty-one-year-old poet, playwright, and philosopher, and officer commanding the local brigade of the IRA, died in 1920 after going seventy-four days without food. Great crowds thronged the streets of London and Cork, to mourn the man who'd proclaimed: "The contest on our side is not one of rivalry or vengeance, but of

endurance. It is not those who can inflict the most, but those that can suffer the most who will conquer . . ."

At the height of The Troubles in the north of Ireland, Prisoners in [notorious prison] Long Kesh were demanding to be treated as prisoners of war, rather than common criminals. They launched a hunger strike in the fall of 1980, and as Christmas approached hunger striker Sean McKenna was near death. It appeared that Margaret Thatcher's government was going to accede to the prisoners' demands, so they called off the strike, but in January 1981, the British government reneged on its concessions and a new hunger strike began.

IRA volunteer Bobby Sands, poet, lover of birds, and nine other prisoners starved to death. Bobby died on May 5, 1981, after going without food for 66 days. He was 27 years old. People mourned the hunger strikers' deaths throughout the world.

MAURA DECIDED UPON HER OWN tactic independently: "I told absolutely nobody about my plans, but I did inform the owners of the *Solitaire*. I had no contact with Shell, never have and never will. That's not good for you. But I told the *Solitaire*'s owners that if the vessel came inside Broadhaven Bay I would have to begin a hunger strike—which would end either when she had left Irish territorial waters, or with my death, whichever.

"And that was it. And then I was sitting in my car, with my handwritten notice ready. The minute Pat O'Donnell told me that *Solitaire* had come into Broadhaven Bay I put up the notice on my car window, stating that I was on hunger strike."

There was a 24-hour-a-day vigil at the pier, and people had Maura's back the entire time.

"I hadn't done any research on hunger strikes," she says now. "I do remember when Bobby Sands and the others died."

Maura wrote to the owners of the *Solitaire*, telling them that as a child she'd suffered from tuberculosis and that it had scarred her lungs. *But how does one conduct a hunger strike? What happens to the human body when you refuse to give it nourishment? How long might she, a very small woman, last without food?*

"It was very instructive in teaching you how to conserve energy," she says. "You didn't run or scuffle about the place. But that was it. No thinking. It was simple—a black or white situation. Either the *Solitaire* was there and I would continue my hunger strike, or else it would remove itself and I could then come off the strike."

A kind neighbor brought Maura a nice large, light, warm blanket, a most practical gift for someone sitting in a car all night beside the sea. A local doctor, receiving an enormous number of calls about Maura's health, came to check on her every day.

The crew of the *Solitaire*, and their bosses, did not attempt to communicate with the hunger striker.

"But then the boat did leave Broadhaven Bay after a few days," Maura says, "and sailed back to Kelly Begs. Because, they claimed, some part of the *Solitaire* was damaged. But it was still in Irish territorial waters., and so I stayed where I was."

On the ninth night, she was getting extra pressure from her supporters to quit the fast. "'Maura,' they said to me, 'you've made your point. You have your moral victory.'

"It was a horrible night," she remembers. "I asked people to stay with me for another night. And yet I still absolutely refused to come off until I could see proof that the vessel was out of our territory.

"And I got it the next day. That was it."

Maura's successful hunger strike had lasted ten full days.

"THE *SOLITAIRE* WAS GONE FOR A YEAR," says Maura. "Then it came sneaking in the second time, and the community was on total lockdown. I tell you it's not nice to be in a place that's locked down by a holy alliance of state and its corporate mercenary forces.

"There were three hundred cops down on the shore, and if anybody from the Solidarity Camp so much as moved they were pounced on. So they got their pipeline laid down in the bay. But it would have been ridiculous to go on a hunger strike again.

"AND HERE WE ARE NOW IN MAYO," Maura says. "The same land is here, and it's the same land out under the sea. It's *our* land. *Irish* land, and I can tell you if Mayo native and social-justice organizer Michael Davitt was around today he wouldn't be giving our country away, like our own government is doing. And it shows you the spineless 'leaders' we have in power today. They wouldn't even think of 1916, or mention the names of people like Michael Davitt.

"Bloody rotten. Every area of the state. I went to school in the 'fifties and 'sixties, a Roman Catholic world, walking to school, enjoying a delightful childhood. What would I call Ireland today? It's a rotten borough. In the old days, people bent the knee and kissed the ring of the Catholic Church. Now, instead of kissing the [eccleasiastical] ring they kiss the corporate ass."

Still, Maura believes her actions and those of all who opposed the Corrib gas project were neither in vain nor without substantial success. "Shell planned to open the refinery by 2003, and didn't finish until December 2013," she notes. "That means we cost them billions. They still have the offshore infrastructure and the state has no control over it because it's all privately owned. We showed, by standing up and hanging in, that the police are a political arm of the state and they protect the interests of those who own everything. Thank heavens few of them have guns.

"One hundred years after 1916, we are worse off than we were then. Servility to the Church followed seamlessly by servility to corporations. That is where we are now."

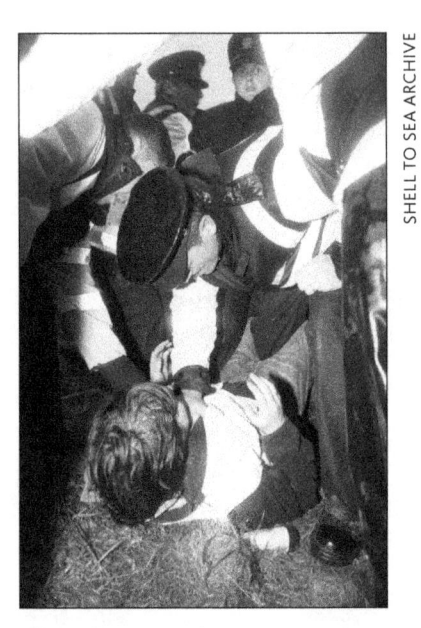

The police at work in County Mayo

Orders

> Everyone will be carrying out orders.
> Where do they come from? Always from elsewhere.
> —R. D. Laing, *The Politics of Experience*

ON NOVEMBER 10, 2006, opponents of Shell Oil's refinery gather at the construction site. They are here to commemorate the eleventh anniversary of the execution of Ken Saro-Wira and eight other nonviolent Nigerian environmental activists. Nine white crosses, each with the name of a victim, stand like witnesses across the road. Next to these crosses, a sign proclaims:

MURDERED BY SHELL, 1995

Irish police do not carry weapons that fire bullets or send violent electric shocks into human flesh. Wearing bright green coats on this brisk fall morning, they appear rather bored, but when ordered to clear the road they wade straight into the crowd, punching and kicking, tossing men and women into ditches. They make no arrests. Four demonstrators are taken to the hospital. Battered and bruised, others limp away.

WATCHING VIDEO COVERAGE of this violence, I flash to Washington, DC, May Day, 1971. Determined to close the capital down for a day in protest of the war on Vietnam, my fellow peace demonstrators and I face a row of squad cars. Sirens wail, helicopters flap overhead, and suddenly without warning the armored police plow straight into us. We flee, only to be stopped by a phalanx of cops firing teargas canisters at our heads and more police blocking our escape, attacking us with nightsticks.

I scramble until, sick and teargas blind, I raise my hands high, shouting "Nonviolence!" A cop smashes my knee to pieces with his nightstick. My leg looks like the result of "a rock tossed through a glass window," say doctors at Georgetown University Hospital.

I will never know who ordered the police to violently attack peaceful antiwar demonstrators that day. Something within us expected the police to join our efforts to end the carnage in Vietnam. Because surely they, too, wanted to bring their fellow Americans home from this unwinnable war.

In 1968, Richard Nixon had claimed he had a secret plan to end the war. He was lying. Before the madness in Southeast Asia ended, twenty thousand more American soldiers, and hundreds of thousands of Vietnamese, would die. Surviving soldiers and civilians would be permanently scarred in one way or another.

SHELL'S OPPONENTS DO NOT HURL insults or objects at the police. Occasionally, someone might lose their temper, push back in a shoving match, grab an officer's hat and toss it away. But they do not punch, or kick, or spit at police.

IN A LITHOGRAPH ON MY LIVING ROOM wall, a soldier sits amid a pile of rubble. He is holding his head in his hands, his feet resting upon the caption *"I was only following orders."*

AT THE NUREMBERG TRIALS FOLLOWING the Second World War, German officials who had helped exterminate millions of human beings argued that they'd been acting as servants of the state. Doing what they were told. Performing their patriotic duty. Following orders. Therefore, they could not be held liable for war crimes.

The International Tribunal disagreed, as reflected in the Nuremberg Principles that resulted from its inquiries:

> Individuals have international duties which transcend the national obligations of obedience imposed by the individual state. He who violates the laws of war cannot obtain immunity while acting in pursuance of the authority of the state if the state in authorizing action moves outside its competence under international law.[1]

The importance of the Nuremberg and subsequent Tokyo trials, asserts Telford Taylor, a key member of the Nuremberg prosecution team, is that they were "landmark developments that planted seeds of new understanding on the part of citizens as to their political obligations. The Nuremberg concept was extended down the ladder of responsibility from the level of primary leaders and applied to doctors, judges, and business executives who were associated with implementing one or another facet of officially sanctioned Nazi (and Japanese imperial) policies."[2]

MY FATHER AND OTHER MEMBERS of my family served in the Second World War, and I grew up watching documentary films and reading books about the war. When we were hunting together in the frozen fields of Iowa, my dad told me strange, disturbing stories about the things human beings do to one another in war.

Later, I learned about horrifying places like Auschwitz and Bergen-Belsen. I tried to imagine anyone forcing children into gas chambers. *What kind of person would kill babies? Who would be willing to starve women and children to death?*

Ordinary men and women, it turns out. Those who are taught by their parents and teachers and clergy, by important, powerful figures, to believe that good (obedient) children must follow orders, that successful people—doctors, lawyers, teachers, soldiers, and prominent politicians—follow orders. That people who refuse to follow orders are unpatriotic, traitors to be cast out, ostracized, sent to prison... or killed.

> Subject to public order and morality, the right of citizens to peaceful assembly without arms is guaranteed by Article 40.6.1.
> —CONSTITUTION OF THE REPUBLIC OF IRELAND

IN THEIR BOOK *AN GARDA SÍOCHÁNA: An Analysis of a Police Force Unfit for Purpose,* Shell opponents Tom Hanahoe, Terence Conway, and John Monaghan write:

> In October 2006, around 200 Gardai—including members of the riot squad—were temporarily posted to police the area, ensuring the highest Gardai in local population ratio, by far, of any rural community in the entire country. The force's objectives were not benign. What ensued was a quite startling erosion of civil liberties, human rights, and democracy. Essentially, democracy was suspended in the region.
> The story of the community's years of suffering and repression, at the hands of Garda Síochána, began with the October 1996 announcement of a natural gas discovery 83 kilometres off Mayo's northwest coast....
> October 2006 marked the beginning of the most intensive, intrusive, prolonged and violent public-order policing operation in a small area in the entire history of the state.[3]

What role, ask these authors, should *the police play in a genuine democracy?* Clearly, as they knew firsthand, during the years Shell Oil was building its refinery in Erris, County Mayo, people felt as though they were living in a "police mini-state," with relentless violence perpetrated against them.

There was a pervasive belief among anti-pipeline protesters that Garda[i] had been given some form of governmental exemption from prosecution and punishment for any repressive or thuggish actions used by them against campaigners. . . .

The pipeline controversy had become a law-and-order issue, rather than a civil liberties or human rights issue. Law-and-order concerns can be, conveniently, concocted to justify even the most absurd and anti-people governmental policies and activities, and also to excuse completely disproportionate responses by a state's security agencies. In Erris, under the pretext of enforcing law and order, people were criminalized for seeking changes to a project that threatened their community's health, safety, and lives. People's rights were ignored. Existing regulations were changed, to meet Shell's needs. Democratic principles were jettisoned.

IN THE EARLY MORNING HOURS of March 16, 1968, soldiers from Charlie Company, 1st Battalion, 20th Infantry Regiment, 11th Brigade, 23rd (Americal) Infantry Division, launch an attack in the Vietnamese village of My Lai, within the larger village of Son My. In the weeks prior to this assault, Charlie Company had suffered several dozen casualties. Angry, frightened soldiers have now been ordered to burn houses, kill livestock, destroy food supplies, and poison wells. There will be few if any civilians in this area, they are told, and people they do find will most likely be communist sympathizers.

The villagers, preparing for market day, do not panic or run away. Then, without warning, young infantrymen start killing them. Before the massacre is over, hundreds of women, children, babies, and older men lie dead. First reports on the slaughter said that 128 Viet Cong and 22 civilians were killed during a "fierce firefight."

On November 12, 1969, investigative journalist Seymour "Sy" Hersh breaks the real My Lai story in the Associated Press. Photographs of babies lying dead on their mother's chests fly around the world. Many people refuse to believe that U.S. soldiers would kill women and children: *Who is to blame for this slaughter? Were Charlie Company's soldiers hardened killers? Were they insane? Who ordered these soldiers to commence this homicidal rampage?*

Of twenty-six men initially charged in the My Lai massacre, Second Lieutenant William Calley Jr. is the only one convicted of premeditated mass murder. Calley's trial lasts four months, and he is sentenced to life in prison. He insists that he had been "only following orders" from his commanding officer. Other members of Charlie Company also had (successfully) used this defense to argue their cases.

William Calley spends three and a half years under house arrest at his apartment in Fort Benning, Georgia, before being freed in 1974 by the Secretary of the Army.

Telford Taylor, the senior U.S. prosecutor at Nuremberg, would write in his book *Nuremberg and Vietnam: An American Tragedy* (1970) that legal principles established at the Nuremberg and Tokyo war crimes tribunals could have been used to try American military commanders who had the duty to prevent mass killings like My Lai.[4]

IN EARLY MAY 1961, SEVEN YEARS before the My Lai massacre, social psychologist and Yale professor Stanley Milgram designs an experiment in which he hopes to learn something about the conflict between obedience to authority and personal conscience. Like people throughout the world, Milgram has been deeply disturbed by the Holocaust genocide of Jews—his own parents were Jewish immigrants to the USA—and mass murders of the Roma people, communists, and others during the Second World War.

Who, he wonders, *actually participated in such terrible crimes? Were some human beings predisposed to harm, even kill others? Or, given the right circumstances, might anyone follow orders to commit murder? How far would people go to obey instruction from an authority when it directly involves harming another person?*

Professor Milgram recruits forty male volunteer "teachers," telling them that the experiment is intended to focus on memory and learning. Each of them will be introduced to a "supervisor/

experimenter" and to a "learner" (both of whom, unknown to the volunteers, are hired actors).

The teacher watches the supervisor attach electrodes to the learner's arms, after which the teacher sits in another room at a console with switches marked in 15-volt increments, ranging from 15 to 450 volts.

The teacher's job is to ask questions of the learner. If the learner is able to answer accurately, it is fine. If he makes a mistake, the teacher must give him a shock. Each mistake requires increased voltage.

The learner does not in fact receive real shocks. Instead, a pre-taped recording is triggered each time a shock switch is pressed. Out of sight behind the screen, the learner complains, whines, shouts that he has a bad heart. He pleads to be released from the experiment. If the teacher pauses, distressed that he might actually be harming the learner, the supervisor/experimenter, wearing a gray lab coat, tells him: "Please go on. The experimenter requires that you go on. It is absolutely essential that you continue. You have no other choice, you must go on."[5]

When a teacher frets about being held responsible for injuring the learner, the experimenter says not to worry. Only he, the individual in charge, will be held accountable.

Prior to Milgram's experiment, experts in human behavior had predicted that only 1 to 3 percent of designated "teachers" would continue giving shocks after a certain point. Only a psychopath or someone with serious psychological problems would follow orders to keep shocking another person.

Although most of them are uncomfortable following the experimenter's orders, all forty of Milgram's volunteer teachers continue shocking the learner up to 300 volts. Sixty-five percent of teachers do not stop administering shocks, and *no one* stops when the learner screams that he is having heart trouble.[6]

Critics have argued that Professor Milgram was untruthful in his dealings with those who volunteered for his experiment. They've also pointed out shortcomings of his methodology. Nevertheless,

Milgram demonstrated that ordinary people are *willing to obey authority figures who order them to harm people who have simply failed to respond correctly to verbal cues.*

LAWRENCE MOSQUEDA, PROFESSOR OF POLITICS and political economy at Evergreen State College, argues that military personnel "have an obligation and a duty to obey lawful orders and indeed have an obligation to disobey unlawful orders, including orders by the President that do not comply with the Uniform Code of Military Justice. The moral and legal obligation is to the U.S. Constitution and not to those who would issue unlawful orders, especially if those orders are in direct violation of the Constitution and the UCMJ."[7]

Under the Nuremberg Principles, which codified guidelines for determining what constitutes a war crime, citizens have not only a right but also an obligation *not* to follow the orders of leaders who are preparing crimes against peace and crimes against humanity.

Ethical people are bound by what U.S. Chief Prosecutor Robert H. Jackson declared in 1948: "The very essence of the [Nuremberg] Charter is that individuals have international duties which transcend the national obligations of obedience imposed by the individual state."[8] The Tokyo War Crimes Trials confirmed and strengthened this mandate, saying that "Anyone with knowledge of illegal activity and an opportunity to do something about it is a potential criminal under international law unless the person takes affirmative measures to prevent commission of the crimes."

SAID ONE SQUAD LEADER WHO participated in the My Lai massacre:

> A lot of people talk about My Lai, and they say, "Well, you know, yeah, but you can't follow an illegal order." Trust me. There is no such thing. Not in the military. If I go into a combat situation and I tell them, "No, I'm not going. I'm not

going to do that. I'm not going to follow that order," they'd put me against the wall and shoot me.⁹

It was not uncommon in the later years of the Vietnam War, however, for soldiers to disobey direct orders, actions for which none were shot by their commanding officers.

ON AUGUST 14, 1971, THREE YEARS after the massacre at My Lai, Stanford University psychology professor Philip Zimbardo initiates the "Stanford Prison Experiment," to measure how people behave when granted power over others.

Funded by the U.S. Office of Naval Research, the experiment is an "investigation into the causes of difficulties between guards and prisoners held by the United States Navy and United States Marine Corps." It is designed to test the hypothesis that prisoners and their guards come into their relationship with existing traits that determine their behavior—be it submissive, abusive, or violent—toward each other.

Zimbardo conducts his experiment in the basement of Stanford's Psychology building, Jordon Hall, in a simulated jail. It will take place over two weeks. Twelve student volunteers are assigned the role of prisoner, twelve other students will act as guards. Prisoners wear uncomfortable ill-fitting smocks, stocking caps, and chains around one ankle, while guards wear khaki shirts and pants, and mirrored sunglasses.

Professor Zimbardo acts as superintendent of a mock prison, with cells holding three prisoners each, a small corridor for a prison yard, and a closet for solitary confinement. Guards are instructed to call student prisoners by their assigned numbers and not allow prisoners privacy in order to take away their sense of individuality and induce a sense of powerlessness.

The results of this experiment are so shocking that it has to be called off after six days. One prisoner appears to go mad after just 36 hours, crying and screaming, flying into a rage. Guards harass prisoners, using a variety of methods to punish them, including protracted exercise, taking away their mattresses, and refusing to allow them to urinate or defecate anywhere but in a bucket in the cell. Guards become increasingly cruel. Experimenters report that one-third of the guards exhibit sadistic tendencies.

Both guards and prisoners in Zimbardo's experiment are Caucasian, middle-class, male college students who have been chosen for their lack of known psychological issues, medical conditions, or criminal history. Their behavior, researchers decide, is a reaction to the situation rather than an expression of individual personality traits.

Using this interpretation, the results are compatible with those of Stanley Milgram's obedience experiments, in which "ordinary" men complied with sadistic orders to electrically shock other people.

The volunteers in each case were "only following orders" and using the power they'd been granted.

"THE RIGHT TO DISSENT PEACEFULLY is a cornerstone of democracy," write Hanahoe, Conway, and Monaghan.

> Civil disobedience and other forms of peaceful protest and dissent, though irksome to governments and The Establishment, are vital for a healthy society," they continue. "Mass dissent is a form of civil activism, an expression of deep-rooted anger in society. The crushing of such dissent suggests authoritarian rule, not democracy. Nevertheless, 'the legitimate use of force in civil society' is widely used by police in Western 'democracies' to curb dissent, to impose harsh and unpopular—even loathsome—government policies, and to maintain and magnify gross inequalities between social classes.
>
> There is, clearly, a considerable estrangement between An Garda Siochána and Irish citizens, and also widespread public distrust of the force. Such estrangement and distrust make the force 'unfit for purpose' in a, purportedly, democratic Irish society. It has been the experience of too many citizens

that their ostensible protectors—their police force—became their oppressors and tormenters, with apparent support of their political masters.[10]

ON MAY 11, 1960, ISRAELI AGENTS track down Adolf Eichmann in Argentina. At his trial for crimes against humanity, war crimes, and crimes against Jews, Eichmann argues that he had been a loyal servant of the German state, following orders from his superiors. Aiding in the murder of millions, he seems to believe, demonstrates a solid sense of character, rather than the behavior of an ideological fanatic.

Psychiatrists who examine Adolf Eichmann find a remarkably "normal" man, not bothered much by guilt. In his essay "A Devout Meditation on Adolf Eichmann" in the 1964 collection *Raids of the Unspeakable,* essayist, social critic, and Catholic monk Thomas Merton writes:

> He was thoughtful, orderly, unimaginative. He had a profound respect for system, for law and order. He was obedient, loyal, a faithful officer of a great state.
>
> Apparently he slept well. He had a good appetite, or so it seems. . . . Eichmann was devoted to duty, self-sacrifice, and proud of his job.
>
> . . . .
>
> I am beginning to realize that "sanity" is no longer a value or an end in itself. The "sanity" of modern man is about as useful to him as the huge bulk and muscles of the dinosaur. If he were a little less "sane"—a little more doubtful, a little more aware of his absurdities and contradictions—perhaps there might be a possibility of his survival. We can no longer assume that because a man is "sane," he is therefore in his "right mind."[11]

DURING THE FIFTEEN-YEAR UPRISING against Shell Oil in

County Mayo, police officers followed orders again and again to attack their fellow citizens. People who were shocked to discover that individuals they knew well were willing to harm them. Some individual police did refuse to monitor protests, even as others were obviously committed to this violence. Still others appear, in videos, to push and shove and swing their batons with a kind of listless energy.

In County Mayo, numerous men, women, and children refused to bow to the power of the state, to succumb to the intimidation of a multinational corporation, to be coerced by the Catholic Church or the police to store their conscience in some safe, excusable place. They refused orders to obey unjust laws, to accept corruption, or to live lives of quiet desperation.

NORWAY'S CENTRAL BANK RECOMMENDED in late 2017 that its "$1 trillion sovereign wealth fund" drop its oil company investments. The Church of England has voted to sell its assets in fossil fuel companies that lack decarbonization strategies aligned with the Paris Climate Accords. Pope Francis sat down with executives of the world's largest oil companies and told them to get out of the business and into renewable energies. To date, scores of Catholic institutions have divested from fossil fuels, heeding the pope's climate encyclical of 2015.

BUILDING ITS PIPELINE AND REFINERY in County Mayo was a Pyrrhic victory for Shell Oil. Now, when the world is experiencing the terrible consequences of global warming, Shell and other multinational companies continue their quest to profit from fossil fuels.

But their time is coming to an end, like the fossils on which they built their fortunes.

The worldwide movement to divest from fossil fuels, to conserve energy, and to implement renewable energy sources and transportation means is growing. The song of life, like the courageous people of Mayo, is impossible to silence.

AS I WRITE THESE WORDS, at the end of July 2019, Ireland has been lauded worldwide for *declaring* a "climate emergency" but no one appears to know exactly how this might translate into *action*.

Will those who continue to profit from polluting and poisoning Mother Earth agree to stop doing so? Will people blinded by greed choose alternative methods of providing energy to the nation they claim to love?

People with whom I have spoken in Ireland want to believe that politicians and their corporate friends will now go beyond rhetorical flourishes and self-serving promises. This will happen only if people from all walks of life, all political persuasions, all religious traditions stand together to create new ways of providing safe, renewable, alternative energy. It can be done. In parts of the world it is being done.

Ireland might turn around, its government wake up to the fact that this tiny island nation, in a time of rising seas and weirding weather, cannot continue to partner with such dinosaurs. Without the determination and courage of a small number of people, Ireland would not have taken even these first steps to become a leader in climate action. The Irish people can still save Ireland.

Acknowledgments

I wish to thank all of the people who helped in many wonderful ways to write this book.

Gerry Bourke, Willie Corduff, Mary Corduff, Mark Garavan, Adelaide Park Gomer, Maura Harrington, William Hederman, Colm Henry, Gabrielle Henry, Mary Horan, Niall King, Paul Lynch, Uinsionn MacGrath, Liamy MacNally, Geradine MacNally, Sr. Majella McCarron, John Monaghan, Brid Ni Seighin, Treasa Ni Ghearraigh, Pat O'Donnell, Eoin O'Leidhin, Micheál Ó Seighin, and Caitlin Ó Seighin.

With special thanks to: Betty Shultz for her hospitality and friendship and to her late husband, Fritz; Peter Wilcock/Redback Photography for the compelling photos; Maura Stephens for her brilliant editing; and Terence Conway, who provided books, articles, and CDs, arranged interviews, and tirelessly drove me about Ireland. Without Terence's and Maura's help, I could not have completed this book.

Glossary

An Garda Síochána. Ireland's national police force. Established in 1922, Gardai do not carry firearms. During the resistance to Shell Oil in County Mayo, Gardai were sent to protests where, apparently following orders, they attacked demonstrators.

An Bord Pleanála (The Planning Board). Quasi-judicial appeals board that rules on infrastructure projects including trans-European energy infrastructure. Its mission, from its website (www.pleanala.ie): "To play our part as an independent national body in an impartial, efficient and open manner, to ensure that physical development and major infrastructure projects in Ireland respect the principles of sustainable development, including the protection of the environment."

Broadhaven Bay. A natural bay of the Atlantic Ocean in western County Mayo. Designated by the National Parks and Wildlife Service as a candidate Special Are of Conservation in 2000. Whales, dolphins, and seals live in the bay. Fishermen take crabs, lobsters, and other kinds of sea life from the bay's waters.

Shell Oil ran its pipeline beneath the bay to landfall, then on to the small town, Ballinaboy, approximately six miles inland.

Michael Collins (1890–1922) was a leader of the Irish Republican Army. He fought for Irish independence in the 1916 Easter Uprising, and was sent to the Frongoch internment camp as a prisoner of war. When released from prison, he managed to outsmart and to kill key British intelligence agents and to plan and lead the Irish War of Independence (aka the Anglo-Irish War). Collins supported the Anglo-Irish Treaty establishing the Irish Free State; he was killed by anti-treaty forces in an ambush on August 22, 1922.

Corrib Gas Project. In 1996, Enterprise Energy Ireland Ltd. and three partners discovered vast quantities of "natural" gas about fifty-two miles off the coast of County Mayo. Several approvals to develop the Corrib Project were issued in 2001, the year before **Royal Dutch Shell** acquired the rights from Enterprise and became lead developer. **Shell E&P Ireland** wanted to develop offshore operations including wells and subsea facilities, offshore pipelines that made land at **Glengad,** and onshore pipelines that would then take it 5.6 miles to an onshore processing plant at **Ballinaboy.** Also in 2002, Minister for the Marine **Frank Fahey,** signed compulsory acquisition orders for access to private lands in and around the village of Rossport for an onshore pipeline route. The following April, **An Bord Pleanála** refused planning permission for the Ballinaboy onshore terminal, slowing down preparations until October 2004, when it approved Shell's new planning application with forty-two conditions. Meanwhile, public opposition was growing. The Rossport 5 were jailed for contempt of court for disrupting pipeline proceedings from June to September 2005. Rossport farmer Willie Corduff was assaulted and hospitalized in April 2009; fisherman Pat O'Donnell, and his crew were rescued after his boat was sunk in June 2009 off Erris Head; An Bord Pleanála declared half of the modified pipeline route "unacceptable" on safety grounds; and international opposition included former U.N. assistant secretary general Dennis Halliday and Archbishop Desmond Tutu. Despite the tremendous public opposition, the pipeline and other infrastructure were completed, and the first gas was pumped ashore just before the new year 2016.

County Mayo. Third largest of Ireland's thirty-two counties. First inhabited by humans during the Mesolithic (Middle Stone) Age around 4500 BC, two millennia after the first nomadic tribes of hunters and fishers arrived in Ireland. It encompasses spectacular coastal "Great Wild Atlantic Way" regions and boasts the highest cliffs in Ireland. The people of the agrarian county were said to be 90 percent dependent on the potato, and therefore the area was especially hard hit by the Great Hunger (1845–52). Motto: *Dia is Muire Linn.* "God and Mary be with us."

Michael Davitt (1846–1906) and The Land League. The Land League was cofounded by Michael Davitt to secure rights for tenants of tenure, fair rents, and freedom to sell property. Born in Straide, County Mayo, Michael and his family were evicted from their home when he was ten years old. He joined the Irish Republican Brotherhood. Arrested in London, accused of sending firearms to Ireland, he spent seven years in prison.

Easter Rebellion, 1916. Armed insurrection in Ireland during Easter Week, launched to drive the British from Ireland and to establish an Irish Republic. Fighting lasted a week, during which the British brought in thousands of troops, heavy artillery, and a gunboat. Most of the rebellion's leaders were executed.

Erris, County Mayo. A barony, Erris encompasses two civil parishes, Kilcommon and Kilmore, on the magnificent Erris Peninsula. The town of Bangor Erris links Belmullet with Ballina and Westport.

European Union. A political union of twenty-eight member-states governing common economic, social, and security policies.

Judge Joseph Finnegan. As President of the High Court he was technically the second-highest ranking judge in the country. During his tenure he presided over sending five of Shell's opponents to prison. He was promoted to the Supreme Court in November 2006 and retired in 2012.

Gombeen Man. A pejorative term used to describe someone who is unscrupulous, shady, willing to cheat people out of their money; used during the Great Hunger to mean a corrupt money-lending "water boy" working for the British.

The Great Hunger (1845–50). Often called "the Irish famine," it began with the potato blight in 1845 that caused the loss of the crop islandwide. One million Irish were forced to leave the country; another million died from disease and hunger. Food that might have fed the starving masses was exported to England.

Irish Civil War (June 28, 1922–May 24, 1923). Following the Irish War of Independence, a conflict broke out between those who supported the Anglo-Irish Treaty establishing the Irish Free State, an entity independent of the United Kingdom but still within the British Empire, and Irish republicans and nationalists who saw this as a betrayal of the Irish Republic proclaimed during the Easter Rebellion. Supporters of the Irish Free State prevailed, leaving bitterness and divisions in Ireland for generations.

Irish Republican Army (IRA, established 1919 as successor to the **Irish Volunteers).** Men and women dedicated to creating a Republic of Ireland. Committed to using armed force to reunite Ireland and, until the Good Friday Agreement of 1998, actively engaged in a war with British troops in the North of Ireland through its Provisionals, or Provos, wing. It would take until July 2005 for the whole IRA—the nonviolent "Officials" wing and the Provos—to finally announce that it was ending its armed campaign and would pursue only peaceful means to achieve reunification of Ireland with Northern Ireland

Irish War of Independence. Popular support for Irish independence from England—Home Rule—grew following the Easter Rebellion of 1916. The republican (independence) party, Sinn Féin, political wing of the IRA, won a landslide victory in 1918, and the IRA began capturing weaponry and freeing republican prisoners. On January 21, 1919, republicans formed a breakaway government, declaring Irish independence. But the British outlawed the new government and also Sinn Féin. Britain brought in the mercenary Black and Tans, who were notorious for their brutal attacks on civilians, to aid the official Royal Irish Constabulary. In May 1921, Ireland was partitioned under British law by the Government of Ireland Act that divided the country into Northern Ireland and the Irish Free State.

Rossport 5, 2005. Three County Mayo farmers and two teachers who challenged Shell Oil's right to violate the rights of property owners that were sent to prison. They spent ninety-four days behind bars, refusing to apologize to the court for their actions, inspiring people in Ireland and elsewhere to join the movement to stop Shell from building the pipeline and gas refinery in rural Ireland.

Shell to Sea. Men, women, and children who resisted Shell Energy's efforts to construct a dangerous gas pipeline and refinery in County Mayo. Among its aims were to ensure safety and not expose the local community to health, safety, and environmental risks; to secure any fuel for use in Ireland; and to seek justice for the human rights abuses to which resisters were subjected. Shell to Sea was an egalitarian group, with no officers and no chairperson. Its strategies were based upon community conversations and general agreement. ShellToSea.com is still being administered as of July 2019.

John Millington Synge (1871–1900). Irish playwright, most famous for penning *Playboy of the Western World*, a humorous play first performed in Dublin at the Abby Theatre on January 26, 1907. It provoked riots there and later in New York City for poking fun at the people of rural Ireland. It continues to be performed—and to spark controversy—more than 100 years later.

Wild Atlantic Way. A stunningly beautiful and, as its name implies, wild tourism itinerary on the west coast, and parts of the north and south coasts, of Ireland. A driving route spanning 1,553 miles, it passes through nine counties, stretching from County Donegal's Inishowen Peninsula, Kinsale, County Cork, on the Celtic Sea. www.thewildatlanticway.com

Notes

INTRODUCTION
1. Shell Energy website, "Communities," www.shell.com/sustainability/communities.html.
2. Shell Global website, "Engaging with Communities," www.shell.com/sustainability/communities/working-with-communities.html
3. Article 40.3, Constitution of Ireland, which can be read in full online at www.gov.ie/en/publication/d5bd8c-constitution-of-ireland.
4. Pope Francis, *Encyclical on Climate Change and Inequality: On Care for Our Common Home* (Brooklyn and London: Melville House, 2015), 15.
5. Global Witness, press release, "2015 Sees Unprecedented Killings of Environmental Activists," www.globalwitness.org/en/press-releases/2015-sees-unprecedented-killings-enviornmental-activists/report.html.

FARMER
1. Bill Poster. "Rossport Five man wins $125,000 Goldman Environmental Prize," indymedia ireland, April 21, 2007. https://www.indymedia.ie/article/82086.
2. Richard Goldman, as quoted in Reuters story "Irish farmer among environmental prize winners," April 21, 2007.

3. "Berta Cáceres, 2015 Goldman Prize Winner: Biography," Goldman Prize, https://www.goldmanprize.org/recipient/berta-caceres/.
4. Global Witness report, "On Dangerous Ground," 2016, executive summary, 4.

FISHERMAN
1. Available at ec.europa.eu/environment/nature/legislation/habitatsdirective/index_en.htm.
2. Shell Energy, "Our Values," www.shell.com/about-us/our-values.html.
3. Shell, "Culture of Transparency," www.shell.com/sustainability/transparency/transparency.html.
4. Article 40.3, Constitution of Ireland.

COMMUNICATOR
1. "Money is the devil's dung, says Pope in fiery speech," *Irish Central*, March 1, 2015. www.irishcentral.com/news/money-is-the-devils-dung-says-pope-francis-in-fiery-speech.
2. Ike Okonta and Oronto Douglas, *Where Vultures Feast: Shell, Human Rights and Oil* (London and New York: Verso, 2003), 96.
3. Lorna Siggins, *Once Upon a Time in the West: The Corrib Gas Controversy* (Ealing, London: Transworld Ireland, 2010), 113.
4. Okonta and Douglas, *Where Vultures Feast*, 44.
5. Oxfam report, "Four EU Countries among World's Worst," December 12, 2016 www.breakingnews.ie/business/oxfam-ireland-is-worlds-6th-worst-corporate-tax-haven-768167.html.
6. Shell, "Who We Are," www.shell/about-us/who-we-are.html.

MUSICIANS
1. Colm and Gabrielle Henry, Closing Remarks to An Bord Pleanála, Oral Hearing into the Corrib Onshore Gas Pipeline Application, September 2010, 2.
2. Colm and Gabrielle Henry, Letter to Planning Board, document shared with author.
3. National Wildlife Federation (NWF) report, "Assault on America: A Decade of Petroleum Company Disaster, Pollution, and Profit," 2010; PDF at www.nwf.org/~/media/PDFs/Global-Warming/Reports/Assault-on-America-A-Decade-of-Petroleum-Company-Disaster.ashx.

4. Maurice and Lorraine McCabe statement as quoted in "Truth today—justice can follow," *Irish Examiner*, February 13, 2017. www.irishexaminer.com/breakingnews/ireland/maurice-and-lorraine-mccabe-statement-truth-today-justice-can-follow-776969.html.
5. Andy Storey and Michael McCaughan. Report for Afri: "The Great Gas Giveaway. How the Elites Have Gambled with Our Health and Wealth," n.d., 1–4, pdf available at www.afri.ie/wp-content/uploads/ 2010/05/The-Great-Gas-Giveaway_2009.pdf.

TEACHER
1. In "They Can Protest, but They Cannot Delay It," August 21, 2006. www.indymedia.ie/article/74215.
2. Maura Harrington, "Reasons for being before Belmullet District Court," document shared with author. September 11, 2013.
3. Global Community Monitor (GCM): this group seems to have vanished online.
4. Brian Barrington for Front Line, "Breakdown in Trust: A Report on the Corrib Gas Dispute," 2010.
5. "Report: Meeting with Special Rapporteur," document provided to the author by Maura Harrington.
6. David Beresford, *Ten Dead Men: The Story of the 1981 Irish Hunger Strike* (New York: Atlantic Monthly Press, 1987), 7–10.

ORDERS
1. The Nuremberg Principles, universally cited and available here: www.tomjoad.org/nuremberg.htm.
2. Telford Taylor, *Nuremberg and Vietnam: An American Tragedy* (Chicago: Quadrangle Books, 1970), 84.
3. Terence Conway, Tom Hanahoe, and John Monaghan. *An Garda Síochána: An Analysis of a Police Force Unfit for Purpose* (County Mayo: June 2014), 4.
4. Telford Taylor, *Nuremberg and Vietnam: An American Tragedy*, (Chicago: Quadrangle Books, 1970), 139.
5., 6. See explorable.com/Stanley Milgram-experiment.
7. Lawrence Mosqueda, *A Duty to Disobey All Unlawful Orders* (Olympia, WA: Evergreen State College, 2003).
8. "Trial of the Major War Criminals, Nuremberg," transcripts, 1945–6. Vol. I, 223–4.

9. "My Lai" episode of *American Experience*, transcript (pdf). www-tc.pbs.org/wgbh/americanexperience/media/pdf/transcript/mylai_transcript.pdf.
10. Conway, 51.
11. Thomas Merton, *Raids of the Unspeakable* (New York: New Directions, 1964), 45, 49.

Related Reading

Beresford, David, *Ten Men Dead: The Story of the 1981 Irish Hunger Strike* (New York: Atlantic Monthly Press, 1987)

Brecher, Jeremy. *Against Doom: A Climate Insurgency Manual* (Oakland, CA: PM Press, 2017)

Clifford, Michael. *A Force for Justice: The Maurice McCabe Story* (Dublin: Hachette Ireland, 2017)

Coogan, Tim Pat. *The Famine Plot* (New York: Palgrave Macmillan, 2012)

Flannery, Tim. *Atmosphere of Hope: Searching for Solutions to the Climate Crisis* (New York: HarperCollins, 2015)

Garavan, Mark. *Our Story: The Rossport 5* (Co. Wicklow, Ireland: Small World Media, 2006)

Hanahoe, Tom. *America Rules* (Dingle, Co. Kerry, Ireland: Brandon, 2002)

Hunt, Timothy J. *The Politics of Bones* (Toronto: McClelland & Stewart, 2005)

Jensen, Derrick. *A Language Older Than Words* (White River Junction, VT: Chelsea Green, 2000)

———. *Listening to the Land* (White River Junction, VT: Chelsea Green, 2002)

Klein, Naomi. *This Changes Everything: Capitalism vs. The Climate* (New York: Simon & Schuster, 2014)

McBay, Aric, Keith Lierre, and Derrick Jensen. *Deep Green Resistance* (New York: Seven Stories Press, 2011)

McCaughan, Michael. *The Price of Our Souls: Gas, Shell, and Ireland* (Dublin: Afri, 2008)

Merton, Thomas. *Raids on the Unspeakable* (New York: New Directions, 1964)

Naidoo, Beverly. *The Other Side of Truth* (New York: Harper Trophy, n.d.)

——. *No Consent: Experiences of Challenging the Corrib Gas Project* (Ireland, 2014)

Ó Mongáin, Séamus, and Treasa Ní Ghearraigh. *Dordán Dulrá: An Introduction to the Natural Landscape of Cill Chomáin in the Barony of Erris, County Mayo* (Ireland: Dún Chaocháin, 2015)

Pope Francis. *Encyclical on Climate Change & Inequality* (Brooklyn/ London: Melville House, 2015)

Russell, Dick. *Horsemen of the Apocalypse* (New York: Skyhorse Publishing, 2017)

Saro-Wiwa, Ken. *Genocide in Nigeria: The Ogoni Tragedy* (London, Lagos/ Port Harcourt: Saros International Publishers, 1992)

Saro-Wiwa, Ken. *Silence Would Be Treason* (Dakar: Council for the Development for Social Science Research in Africa, 2013)

Shah, Sonia. *Crude: The Story of Oil* (New York: Seven Stories Press, 2006)

Shell to Sea. *Liquid Assets* (Dublin: Shell to Sea, 2012)

Siggins, Loren. *Once Upon a Time in the West: The Corrib Gas Controversy* (London: Transworld Ireland, 2010)

Wohlleben, Peter. *The Hidden Life of Trees* (Vancouver/Berkeley: Greystone Books, 2016)